電気・電子を説明す
英語

宮野 晃 著

森北出版株式会社

ま え が き

情報化は近年めざましいものがあり，現在は高度情報化社会ともいわれている．最近では“インターネット”がさまざまなメディアで取り上げられており，多くの人の注目を浴びている．個人レベルでの情報の受信・発信を自由に行うことができる基盤ができつつあるわけだが，そこでも英語という壁が存在しているようだ．英語に自信がないため，インターネットに積極的になれないという声も聞こえてくるからである．

一方，バブルの崩壊やリストラなどが取り沙汰され，また情報整理や勉強法などの書物がベストセラーとなっている現状を考えると，自分の将来に危機感を募らせ，その対抗手段として能力向上を図っている人が増えているということが見て取れそうである．英語力はそのための強力なツールである．文化系，理科系を問わず，英語力の必要性はますます増大していくことは間違いない．

英文の読み書きの力は，コミュニケーション能力の重要な一部をなしている．技術英語に限っていえば，学校で正課として学んだという経験をほとんどの人はもたないはずである．いきおい，社会人となってから必要に迫られて独学するはめになる．英語が苦手なばかりに，必要な情報を手に入れられなかった，あるいは発表できなかったという苦い経験をもつ人も多いのではないだろうか．だが，技術英語といっても，何も特別な英語ではなく，学校で学んだ文法はそのまま役立てることができる．それでも技術英語には，それなりの“クセ”があるのもまた事実である．その特徴を２点ばかり挙げてみると，

① 英文の構造が比較的単純である

② 専門用語が多出する

となるが，特に技術英語は一つの文がいくつもの意味にとれるような文はよくないので，文学的な表現やあいまいな文は避けなくてはならない．「明確に（clear），正しく（correct），簡潔に（concise）」の「３Ｃ」が重要である．また，その分野の知識がないと専門用語を正しく訳せないこともあるので，技術分野の勉強も忘れてはならない．

本書は主に電気・電子工学の分野に携わっている人が，論文やマニュアル類を読んだり書いたりする際に参考になるようまとめたものである．しかし，動詞そのものは特定の分野に限定されているわけではないので，科学技術の

どんな分野の人々にも参考になるであろう．動詞は英文の中心的役割を果たす品詞であり，それゆえに動詞の使い方を身につけることは文書の作成においてたいへん重要なことである．本書ではなるべく簡潔な英文を採用し，和訳文もできるだけ原文に忠実にしたので，読者に不要な混乱をもたらさないはずである．用例やポイントを参考にして，readingやwritingの際に大いに活用していただきたい．本書では使用頻度の高い276個の動詞を取り上げている．採用したかった動詞は他にもあるが，後日,読編としてまとめられたら，と思っている．

本書を手にすることにより，実際に通用する技術英文の姿やレベルを読者が理解し，英語力向上のお役に立つことができれば，著者としてはこの上ない幸せである．

本書は，1996年5月に工業調査会から出版されたものを継続して森北出版より発行することになったものである．

2011年7月　　　　宮野　晃

＊本文中の略語について

［反］反対語　　［類］類語　　n:名詞　　adj:形容詞　　adv:副詞

英字目次

A

absorb ……41
（液体・音・光などを）吸収する

accelerate ……41
（物質などを）加速する／（反応などを）促進する

accomplish ……42
（仕事などを）成し遂げる，達成する

achieve ……43
（仕事・目的などを）成し遂げる，達成する

act ……44
（機械などが）作動する，働く／（力などが）作用する

adapt ……44
（状況などに）適合させる,適応する,合わせる

add ……45
（他の物に）加える,つけ足す

address ……46
（仕事・問題などに）取り組む／（問題などを）扱う

adjust ……46
（機械・装置・電気量などを）調節する,調整する

adopt ……47
（理論・技術・方式などを）採用する,使用する

affect ……48
（物・事に）影響を及ぼす

agree ……49
（物・事が）一致する

allow ……49
～を許す,～を許容する,～を可能にする,～を考慮に入れる

amplify ……50
（電気量・信号などを）増幅する

analyze ……51
（物・事を）解析する,分析する

anneal ……………………………………………………………52
(金属などを) 焼きなます,アニールする

appear ……………………………………………………………52
(物・事が) 現れる,生ずる,～のように思われる (見える)

apply ……………………………………………………………53
(方法・規則などを) 適用する,応用する,使用する／(力・熱・電圧などを) 加える,印加する／(方法・規則などが) あてはまる

approach ……………………………………………………………55
(数量・状態・場所などに) 近づく,接近する／(問題・研究などに) 取り組む／(物・事が) 近づく,接近する

arise ……………………………………………………………56
(問題・困難などが) 生じる,起こる,発生する

assess ……………………………………………………………56
(価値・性質などを) 評価する,算定する,決定する

assist ……………………………………………………………57
(人・物事を) 助ける,支援する,援助する／(人・物事が) 助ける,支援する,援助する

assume ……………………………………………………………58
(～だと) 仮定する,考える,見なす

attain ……………………………………………………………59
(目的・望みなどを) 達成する,成し遂げる,得る

attribute ……………………………………………………………59
(物・事が) ～にあると考える,～に帰する,～のせいにする

average ……………………………………………………………60
(数量などを) 平均する,平均して～となる

avoid ……………………………………………………………61
(望ましくない物・事を) 避ける,回避する

B

become ……………………………………………………………62
(物・事が) ～になる

begin ……………………………………………………………63
(事を) 始める／(事が) 始まる

behave ……………………………………………………………64
(～に対して) ふるまう,挙動する,動作する

bring ……………………………………………………………65
(物を) 持ってくる／(事を) もたらす

build ……………………………………………………………66
(装置などを) 作る,組み立てる／(建築・構造物などを) 建設する,建てる

build up ……66
増大する，増加する

C

calculate ……67
（数式などを）計算する

calibrate ……68
（計測器を）校正する

call ……69
（名前などを）呼ぶ，名づける／（人が）呼ぶ

cancel ……70
（物・事を）取り消す，相殺する

carry ……70
（物を）携行する

cause ……71
（物・事が～の）原因となる／（結果として～を）引き起こす，もたらす

change ……72
（物・事を）変える，変更する／（物・事が）変わる，変化する

characterize ……73
（物・事を）特徴づける

check ……74
（物・事を）調べる，点検する，確かめる，照合する

choose ……75
（複数個の中から～を）選ぶ，選択する

clarify ……76
（考え方・理論などを）明らかにする，明確にする

classify ……77
（物・事を）分類する

coat ……78
（塗料などを）塗る，塗布する，被覆する

coincide ……78
（性質などが）一致する

collect ……79
（信号・データなどを）集める，収集する／（信号・データなどが）集まる，たまる

combine ……80
（物・事を）結合させる，組み合わせる

come ……80
（物が）来る，着く，達する，現れる／（事が）起こる

communicate ……81
（情報・データなどを）伝達する／（人・物が）通信する

compare ……82
（物・事と）比較する

compensate ……83
（損失などを）補償する,補う

compose ……84
（物を）組み立てる,構成する

compress ……85
（信号・データ・気体などを）圧縮する

compute ……86
（数・量などを）計算する

concern ……87
（物・事が）～に関係がある,～に関する,～にかかわる

conduct ……88
（研究・実験などを）行う,実施する

confirm ……88
（証拠・予測などを）確かめる,立証する,確認する

connect ……89
（二つ以上の物を）つなぐ,接続する,連結する／（二つ以上の物が）つながる，接続する，連結する

consider ……90
（問題などを）よく考える,考察する

consist ……91
（部分・要素から）なる

construct ……92
（部品などを）組み立てる

consume ……93
（電力などを）消費する

contain ……94
（物を）含む

control ……95
（物・事を）制御する

converge ……96
（級数・アルゴリズムなどが）収束する

convert ……96
（信号などを）変換する

cool ……98
（物を）冷却する,冷やす

correct ……98
（誤りなどを）訂正する,修正する／（装置・回路などを）補正する

correspond ……99
（物・事が）一致する,相当する

couple ……100
（物を）結合する,連結する

cover ……100
（物を）覆う／（問題などを）扱う／（範囲を）カバーする

D

deal ……101
（～を）扱う,処理する,論じる

decrease ……102
（数・量などを）減少させる,減らす／（数・量などが）減少する,減る

define ……103
（語などを）定義する

degrade ……104
（数・量などを）減少させる／（性能・質などを）低下させる,劣化させる

demodulate ……105
（変調波を）復調する

demonstrate ……105
（学説・真理などを）実証する,証明する,論証する

depend ……106
～に依存する,～に左右される,～によって決まる

deposit ……107
（薄膜などを）堆積させる,蒸着させる

derive ……108
（性質・知識・利益などを）引き出す,導き出す／（物・事の）由来をたずねる

describe ……109
（物・事について）述べる,記述する

design ……110
（物を）設計する

detect ……111
（信号・欠陥・化学物質などを）検出する

determine ……112
（物・事を）決定する／（位置・形状などを）測定する

develop ……113
（物を）開発する

differ ……114
（物・事が）異なる

diffuse ……115
（ガス・液体・熱・光などを）拡散する／（ガス・液体・熱・光などが）拡散する

digitize ……115
（信号などを）ディジタル化する，計数化する

discuss ……116
（物・事について）論じる，議論する

display ……117
（文字・図形などを）表示する

distinguish ……118
（物・事を）区別する，識別する／（物・事の間の相違を）見分ける，区別する

distort ……118
（物体・信号・波形を）歪ませる，変形させる

divide ……119
（物・事を）分割する，分ける

do ……120
（実験・解析などを）行う，する

draw ……121
（結論などを）導く，得る／（線などを）引く，描く

drive ……122
（回路・機械などを）駆動する

E

eliminate ……123
（物・事を）除く，除去する

emit ……124
（熱・光・音・ガスなどを）放つ，放出する，放射する

employ ……125
（物・手段などを）使用する，用いる，利用する

enable ……126
（事を）可能にする

enhance ……126
（価値・性能などを）高める，増大する

ensure ……127
（事を）保証する，確実にする

equal ……128
（数量・大きさなどの点で）等しい

equip ……129
（物を）備えつける,装備する

establish ……129
（理論・学説などを）立証する,証明する／（値などを）確定する,決める／（規則・制度などを）確立する

estimate ……130
（能力・特性などを）評価する,推定する

evaluate ……131
（価値・能力などを）評価する／（数式などの）値を求める

evaporate ……132
（薄膜・層などを）蒸着させる

examine ……133
（物・事を）検査する,検討する,調査する

excite ……134
（電磁石・コイルなどを）励磁する／（アンテナなどを）励振する

execute ……134
（計画などを）実行する,実施する,遂行する

exhibit ……135
（特性などを）示す

exist ……136
（物が）存在する,実在する

expect ……136
（物・事を）期待する,予想する,予期する

exploit ……137
（特性などを）活用する,利用する

explore ……138
（可能性などを）調べる,探る,調査する

expose ……139
（光・放射線などに）さらす

express ……140
（事を）表現する,示す,表す

extend ……141
（範囲・方法・意味などを）拡張する，拡大する

extract ……142
（信号・情報・データなどを）抽出する

F

fabricate …… 143
（装置・機械・製品などを）組み立てる，製作する

facilitate …… 144
（事を）容易にする，促進する

fail …… 145
（機械などが）故障する，働かなくなる／（計画などが）失敗する

fail to do …… 145
～できない，～しそこなう

feed …… 146
（信号・電力などを）送る，供給する，給電する

fill …… 147
（物を）満たす，詰め込む

filter …… 148
（信号などを）ろ波する

find …… 148
（物・事を）見つける，見いだす，発見する，求める

flow …… 150
（電流・液体などが）流れる

focus …… 150
（光などが）集束する／（注意・関心などを）集中させる／（注意・関心などが）集中する

form …… 151
（物を）作る，形成する

formulate …… 152
（事を）定式化する，公式化する，式で示す

G

generalize …… 153
（事実・知識などを）一般化する

generate …… 154
（電気・熱・信号などを）発生させる，作り出す，生み出す

give …… 155
（物・事を）与える，示す，施す

grow …… 156
（薄膜・結晶などを）成長させる／（物・事が）発展する，増大する，成長する

guarantee……157
(物・事を) 保証する

H

handle……158
(事を) 処理する

have……159
(物を) もっている,所有する/ (性質・特性などを) 有する

heat……160
(物を) 加熱する,熱する

help……161
(物・事を) 助ける

hold……162
(物を) 保持する/ (物・事をある状態のまま) 保つ/ (データ・情報などを) 保存 (記憶,格納) する/ (式などが) 成り立つ

I

identify……163
(物・事を) 識別する,確認する,同定する

ignore……164
(物・事を) 無視する

illustrate……165
(物・事を) 示す,例示する,図示する

implement……165
(回路・システムなどを) 実現する,実装する/ (計画・プロジェクトなどを) 実行する,実施する

improve……167
(物・事を) 改善する,改良する/ (性能・特性・価値などを) 高める,向上させる

include……168
(物・事を) 含む,内蔵する

incorporate……168
(物を) 組み込む,含む

increase……169
(大きさ・数量・程度などを) 増やす,増加させる/ (能力などを) 向上させる/ (大きさ・数量・程度などが) 増える,増大する

indicate……170
(物・事を) 示す,表す

induce ……171
（電流・磁気などを）誘導する

influence ……172
（物・事に）影響を及ぼす

integrate ……172
（回路などを）集積（化）する／（物を）まとめる,一体化する,統合する

interact ……174
（物と）相互作用する

interfere ……175
（物・事を）妨害する,干渉する

investigate ……175
（物・事を）調べる,研究する,調査する

involve ……177
（物・事を）含む,伴う

isolate ……177
（回路などを）絶縁する／（物を）分離する,隔離する

K

keep ……178
（ある動作・状態を）続ける,維持する,保つ／（物・事が）～し続ける

know ……179
（物・事を）知っている,わかっている

L

lead ……181
（物・事が）至る,通じる

lie ……182
（事実・理由・欠陥などが）ある

limit ……182
（物・事を）制限する

list ……184
（物・事を）一覧表にする,列挙する

lower ……184
（速度・温度・程度などを）下げる,低下させる

m

maintain ……185
（状態・温度・圧力・性能などを）維持する,保持する

make ……186
（物を）作る,製作する／（物・事を）～にする／（物・事が）～させる

match……189
（物・事と）合う,一致する,適合する／（回路などを）整合する

measure……191
（長さ・量・大きさなどを）測定する／（～の）寸法をしている

minimize ……192
（数量などを）最小にする

model……193
（物・事を）モデル化する

modify ……194
（部分的に）修正する,変更する,改善する

modulate ……195
（周波数・振幅などを）変調する

monitor……196
（状態などを）監視する

mount……197
（物を）取り付ける,実装する,搭載する

move ……198
（物を）動かす,移動させる／（物が）動く,移動する

n

need ……200
（物・事を）必要とする

neglect……201
（物・事を）無視する

note……202
（物・事に）注意する,注目する,気づく,言及する

o

observe ……203
（物・事を）観測する,観察する,認める

obtain ……204
(物・事を) 得る,獲得する

occur ……204
(物・事が) 起こる,生じる,現れる,存在する

offer ……205
(物・事を) 提供する,与える,示す

operate ……206
(機械・装置などを) 動作させる,操作する,運転する／(機械・装置などが) 動作する,作動する,機能する,動く

optimize ……207
(物・事を) 最適化する

overcome ……208
(障害・困難・問題などに) 打ち勝つ／(障害・困難・問題などを) 克服する

P

pass ……209
(物を) 通過させる,通す／(物が) 通過する,通る

perform ……210
(仕事・試験などを) 行う,する／(機械などが) 機能する,動作する

pick ……211
(物・事を) 選ぶ,選択する

place ……212
(物を) 置く,配置する,設置する

play ……213
(～の役割を) 果す

plot ……213
(図形などを) プロットする

polarize ……214
(光を) 偏光させる

predict ……215
(事を) 予測する

prepare ……216
(物・事を) 用意する／(試料などを) 作製する

present ……217
(論文・報告書などを) 提出する,示す

presume ……217
(事を) 推定する,仮定する

prevent ……218
（物・事を）妨げる，防ぐ

process ……219
（信号・データなどを）処理する

produce ……220
（物・事を）作り出す，生み出す，発生させる，引き起こす

propagate ……221
（音・光・電磁波・熱などを）伝える，伝播させる，伝搬させる／
（音・光・電磁波・熱などが）伝わる，伝播する，伝搬する

propose ……222
（事を）提案する

protect ……223
（損傷・危険などから）守る，保護する

prove ……223
（事を）証明する，立証する／（事が）わかる，判明する

provide ……225
（物を）供給する，与える，備える

Q

quantize ……227
（信号・エネルギーなどを）量子化する

R

radiate ……228
（電磁波・光・熱などを）放射する

range ……229
（範囲・程度などが）及ぶ，わたる

reach ……229
（ある状態・結果などに）達する

react ……230
（物・事が）反応する，対応する

realize ……231
（物・事を）実現する

receive ……231
（信号などを）受信する／（注目・好意などを）受ける

record ……232
（物・事を）記録する

reduce ……233
（数量・程度・範囲などを）減らす，減少させる，低減させる／
（ある物・状態などに）変える

reflect ……234
（光・音・熱などを）反射する

relate ……235
（物・事を）関係づける／（物・事が）関係がある

remain ……236
（物・事が）とどまる，～のままでいる

remove ……236
（不要なものを）取り除く，除去する

repeat ……237
（動作・行為などを）繰り返す

replace ……238
（物・事に）取って代わる

report ……239
（研究・調査などを）報告する

represent ……240
（物・事を）表す，表現する

require ……241
（物・事を）必要とする，要求する

restrict ……242
（事を）制限する，限定する

result ……243
（物・事が）生じる

reveal ……244
（ある事を）明らかにする，示す

run ……245
（プログラムを）実行する／（機械･装置などを）動かす／（プログラム・機械・装置などが）動作する／（ケーブル・配線などが）通る

S

sample ……246
（信号などを）標本化する，サンプリングする

satisfy ……247
（要件・条件などを）満たす，満足させる

scatter ……248
（電磁波・光などを）散乱させる

see ……249
（事が）わかる

select ……249
（物・事を）選択する

send……251
（信号・情報などを）送る,送信する

sense……251
（状態・変化などを）検知する,感知する

separate……253
（物を）分離する,分ける

set……254
（物・事を）設定する,セットする

shift……255
（物・事を）移す,移動させる／（ビット列・文字列などを）桁送りする

show……255
（物・事を）示す,表す

simplify……256
（物・事を）単純化する,簡単にする,簡略化する

simulate……257
（特性・機能・挙動などを）シミュレートする

solve……258
（問題・数式などを）解く

stabilize……259
（物・事を）安定化する,安定させる

start……260
（事を）始める／（事が）始まる

state……261
（事実・意見・問題などについて）述べる,示す

store……262
（データを）格納する,蓄積する,保管する,記憶する／（エネルギーなどを）貯蔵する

study……263
（物・事を）研究する,調べる

suffer……264
（損害・作用などを）受ける,被る／（物・事で）損害を被る

suggest……265
（計画・考えなどを）示唆する,提案する

suit……266
（物・事に）適する,適合する

summarize……266
（物・事を）要約する

suppose……267
（事を）仮定する

suppress ……268
（振動・雑音などを）抑制する,除去する,低減する

switch ……269
（装置などの）スイッチを切り換える

synchronize ……270
（回路・信号などを）同期する,同調する

synthesize ……271
（回路・信号などを）合成する

T

take ……272
（時間を）費やす／（注意・決心・見方などを）する／（データなどを）とる／（形状・性質などを）とる

tend ……274
（物・事が）～する傾向がある／（物・事がある結果・状態に）向かう,至る

test ……275
（物・事を）試験する

transduce ……276
（エネルギー・信号などを）変換する

transfer ……276
（データなどを）転送する／（物を）移動させる,移す

transform ……278
（物を）変換する,変形する

transmit ……278
（光・熱などを）伝える／（電波・信号・データなどを）送信する,伝送する

travel ……279
（光・音などが）伝わる,移動する

treat ……280
（物・事を）扱う,処理する,論じる

try ……281
（物・事を）試す,試みる

tune ……282
（ある値に）調整する／（受信機・レーザなどを）同調させる

U

understand ……283
（事を）理解する

use ……284
（物・事を）使う,使用する,利用する

utilize ……286
（物・事を）利用する,役立たせる

V

vary ……287
（物・事を）変える,変化させる／（物・事が）変わる,変化する

verify ……288
（事を）立証する,実証する,検証する,確かめる

view ……289
（物を）見る,観察する

W

work ……290
（機械などが）動作する,作動する,機能する／（計画・方法などが）うまくいく

write ……291
（数式・プログラムなどを）書く

Y

yield ……293
（物・事を）生じる,もたらす／（物・事が）得られる

和語目次

あ行

（物・事と）合う　match ……189
（考え方・理論などを）明らかにする　clarify ……176
（ある事を）明らかにする　reveal ……244
（物・事を）与える　give ……155
（物・事を）与える　offer ……205
（物を）与える　provide ……225
（問題などを）扱う　address ……46
（～を）扱う　deal ……101
（問題などを）扱う　cover ……100
（物・事を）扱う　treat ……280
（信号・データ・気体などを）圧縮する　compress ……85
（信号・データなどが）集まる　collect ……79
（信号・データなどを）集める　collect ……79
（方法・規則などが）あてはまる　apply ……53
アニールする　anneal ……52
（事を）表す　express ……140
（物・事を）表す　indicate ……170
（物・事を）表す　represent ……240
（物・事を）表す　show ……255
（物・事が）現れる　appear ……52
（物が）現れる　come ……80
（物・事が）現れる　occur ……204
（事実・理由・欠陥などが）ある　lie ……182
（物・事が）～にあると考える　attribute ……59
（状況などに）合わせる　adapt ……44
（物・事を）安定化する　stabilize ……259

（物・事を）安定させる　stabilize ……259
（ある動作・状態を）維持する　keep ……178
（状態・温度・圧力・性能などを）維持する　maintain ……185
～に依存する　depend ……106
（物・事が）至る　lead ……181
（物・事がある結果・状態に）至る　tend ……274
（物・事を）一覧表にする　list ……184
（物を）一体化する　integrate ……172
（性質などが）一致する　coincide ……78
（物・事が）一致する　agree ……49
（物・事が）一致する　correspond ……99
（物・事と）一致する　match ……189
（事実・知識などを）一般化する　generalize ……153
（物を）移動させる　move ……198
（物・事を）移動させる　shift ……255
（物を）移動させる　transfer ……276
（物が）移動する　move ……198
（光・音などが）移動する　travel ……279
（力・熱・電圧などを）印加する　apply ……53
（注目・好意などを）受ける　receive ……231
（損害・作用などを）受ける　suffer ……264
（物を）動かす　move ……198
（機械・装置などを）動かす　run ……245
（物が）動く　move ……198
（機械・装置などが）動く　operate ……206
（障害・困難・問題などに）打ち勝つ　overcome ……208
（物・事を）移す　shift ……255
（物を）移す　transfer ……276
（計画・方法などが）うまくいく　work ……290
（物・事を）生み出す　produce ……220
（電気・熱・信号などを）生み出す　generate ……154
（機械・装置などを）運転する　operate ……206
（物・事に）影響を及ぼす　affect ……48
（物・事に）影響を及ぼす　influence ……172

（線などを）描く　draw……121
（複数個の中から～を）選ぶ　choose……75
（物・事を）選ぶ　pick……211
（物・事が）得られる　yield……293
（目的・望みなどを）得る　attain……59
（結論などを）得る　draw……121
（物・事を）得る　obtain……204
（人・物事を）援助する　assist……57
（人・物事が）援助する　assist……57
（方法・規則などを）応用する　apply……53
（物を）覆う　cover……100
（損失などを）補う　compensate……83
（物を）置く　place……212
（信号・電力などを）送る　feed……146
（信号・情報などを）送る　send……251
（研究・実験などを）行う　conduct……88
（実験・解析などを）行う　do……120
（仕事・試験などを）行う　perform……210
（問題・困難などが）起こる　arise……56
（事が）起こる　come……80
（物・事が）起こる　occur……204
（物・事が）～のように思われる（見える）　appear……52
（範囲・程度などが）及ぶ　range……229

か行

（物・事を）解析する　analyze……51
（物・事を）改善する　improve……167
（部分的に）改善する　modify……194
（物を）開発する　develop……113
（望ましくない物・事を）回避する　avoid……61
（物・事を）改良する　improve……167
（物・事を）変える　change……72
（ある物・状態などに）変える　reduce……233
（物・事を）変える　vary……287

（物・事が）～にかかわる　concern ……87
（数式・プログラムなどを）書く　write ……291
（ガス・液体・熱・光などを）拡散する　diffuse ……115
（ガス・液体・熱・光などが）拡散する　diffuse ……115
（事を）確実にする　ensure ……127
（範囲・方法・意味などを）拡大する　extend ……141
（範囲・方法・意味などを）拡張する　extend ……141
（値などを）確定する　establish ……129
（物・事を）獲得する　obtain ……204
（証拠・予測などを）確認する　confirm ……88
（物・事を）確認する　identify ……163
（データを）格納する　store ……261
（物を）隔離する　isolate ……177
（規則・制度などを）確立する　establish ……129
（物質などを）加速する　accelerate ……41
（特性などを）活用する　exploit ……137
（～だと）仮定する　assume ……58
（事を）仮定する　presume ……217
（事を）仮定する　suppose ……267
（物を）加熱する　heat ……160
～を可能にする　allow ……49
（事を）可能にする　enable ……126
（範囲を）カバーする　cover ……100
（物・事が）変わる　change ……72
（物・事が）変わる　vary ……287
（～だと）考える　assume ……58
（問題などを）よく考える　consider ……90
（物・事が）～に関係がある　concern ……87
（物・事が）関係がある　relate ……235
（物・事を）関係づける　relate ……235
（物・事を）観察する　observe ……203
（物を）観察する　view ……289
（状態などを）監視する　monitor ……196
（物・事を）干渉する　interfere ……175

（物・事が）～に関する　concern……87
（物・事を）観測する　observe……203
（物・事を）簡単にする　simplify……256
（状態・変化などを）感知する　sense……251
（物・事を）簡略化する　simplify……256
（データを）記憶する　store……261
（物・事について）記述する　describe……109
（物・事が）～に帰する　attribute……59
（物・事を）期待する　expect……136
（物・事に）気づく　note……202
（機械・装置などが）機能する　operate……206
（機械などが）機能する　perform……210
（機械などが）機能する　work……290
～によって決まる　depend……106
（値などを）決める　establish……129
（液体・音・光などを）吸収する　absorb……41
（信号・電力などを）給電する　feed……146
（信号・電力などを）供給する　feed……146
（物を）供給する　provide……225
（～に対して）挙動する　behave……64
～を許容する　allow……49
（装置などの）スイッチを切り換える　switch……269
（物・事を）記録する　record……232
（物・事について）議論する　discuss……116
（回路・機械などを）駆動する　drive……122
（物・事を）区別する　distinguish……118
（物・事の間の相違を）区別する　distinguish……118
（物・事を）組み合わせる　combine……80
（物を）組み込む　incorporate……168
（装置などを）組み立てる　build……66
（物を）組み立てる　compose……84
（部品などを）組み立てる　construct……92
（装置・機械・製品などを）組み立てる　fabricate……143
（動作・行為などを）繰り返す　repeat……237

（物が）来る　come ……80
（他の物に）加える　add ……45
（力・熱・電圧などを）加える　apply ……53
（物を）携行する　carry ……70
（物・事が）～する傾向がある　tend ……274
（数式などを）計算する　calculate ……67
（数・量などを）計算する　compute ……86
（信号などを）計数化する　digitize ……115
（物を）形成する　form ……151
（ビット列・文字列などを）桁送りする　shift ……255
（価値・性質などを）決定する　assess ……56
（物・事を）決定する　determine ……112
（物・事を）結合させる　combine ……80
（物を）結合する　couple ……100
（物・事が～の）原因となる　cause ……71
（物・事を）研究する　investigate ……175
（物・事を）研究する　study ……263
（物・事に）言及する　note ……202
（物・事を）検査する　examine ……133
（信号・欠陥・化学物質などを）検出する　detect ……111
（数・量などを）減少させる　decrease ……102
（数・量などを）減少させる　degrade ……104
（数量・程度・範囲などを）減少させる　reduce ……233
（事を）検証する　verify ……288
（数・量などが）減少する　decrease ……102
（建築・構造物などを）建設する　build ……66
（状態・変化などを）検知する　sense ……251
（事を）限定する　restrict ……242
（物・事を）検討する　examine ……133
（問題などを）考察する　consider ……90
（事を）公式化する　formulate ……152
（性能・特性・価値などを）向上させる　imorove ……167
（能力などを）向上させる　increase ……169
（物・事で）損害を被る　suffer ……264

（計測器を）校正する calibrate ……68
（回路・信号などを）合成する synthesize ……271
（物を）構成する compose ……84
（損害・作用などを）被る suffer……264
〜を考慮に入れる allow ……49
（障害・困難・問題などを）克服する overcome ……208
（物・事を）試みる try ……281
（機械などが）故障する fail ……145
（物・事が）異なる differ ……114

さ行

（数量などを）最小にする minimize ……192
（物・事を）最適化する optimize ……207
（理論・技術・方式などを）採用する adopt……47
（試料などを）作製する prepare ……216
（可能性などを）探る explore ……138
（望ましくない物・事を）避ける avoid ……61
（速度・温度・程度などを）下げる lower……184
（物・事が）〜させる make ……186
（機械などが）作動する act ……44
（機械・装置などが）作動する operate ……206
（機械などが）作動する work ……290
（物・事を）妨げる prevent……218
〜に左右される depend ……106
（力などが）作用する act ……44
（光・放射線などに）さらす expose……139
（価値・性質などを）算定する assess ……56
（信号などを）サンプリングする sample ……246
（電磁波・光などを）散乱させる scatter ……248
（人・物事を）支援する assist ……57
（人・物事が）支援する assist ……57
（事を）式で示す formulate ……152
（物・事を）識別する distinguish……118
（物・事を）識別する identify……163

（物・事を）試験する test ……257
（計画・考えなどを）示唆する suggest ……265
～しそこなう fail to do ……145
（回路・システムなどを）実現する implement ……165
（物・事を）実現する realize ……231
（計画などを）実行する execute ……134
（プログラムを）実行する run ……245
（計画・プロジェクトなどを）実行する implement ……165
（物が）実在する exist ……136
（研究・実験などを）実施する conduct ……88
（計画などを）実施する execute ……134
（計画・プロジェクトなどを）実施する implement ……165
（学説・真理などを）実証する demonstrate ……105
（事を）実証する verify ……288
（物を）実装する mount ……197
（回路・システムなどを）実装する implement ……165
（物・事を）知っている know ……179
（計画などが）失敗する fail ……145
（特性・機能・挙動などを）シミュレートする simulate 257
（特性などを）示す exhibit ……135
（事を）示す express ……140
（物・事を）示す give ……155
（物・事を）示す illustrate ……165
（物・事を）示す indicate ……170
（物・事を）示す offer ……205
（論文・報告書などを）示す present ……217
（ある事を）示す reveal ……244
（物・事を）示す show ……255
（事実・意見・問題などについて）示す state ……261
（信号・データなどを）収集する collect ……79
（部分的に）修正する modify ……194
（誤りなどを）修正する correct ……98
（回路などを）集積（化）する integrate ……172
（光などが）集束する focus ……150

（級数・アルゴリズムなどが）収束する converge ………96
（薄膜などを）蒸着させる deposit ………107
（薄膜・層などを）蒸着させる evaporate ………132
（注意・関心などを）集中させる focus ………150
（注意・関心などが）集中する focus ………150
（信号などを）受信する receive ………231
（物・事を）照合する check ………74
（問題・困難などが）生じる arise ………56
（物・事が）生じる occur ………204
（物・事が）生じる result ………243
（物・事を）生じる yield ………293
（理論・技術・方式などを）使用する adopt ………47
（方法・規則などを）使用する apply ………53
（物・手段などを）使用する employ ………125
（物・事を）使用する use ………284
（物・事が）生ずる appear ………52
（電力などを）消費する consume ………93
（学説・真理などを）証明する demonstrate ………105
（理論・学説などを）証明する establish ………129
（事を）証明する prove ………223
（不要なものを）除去する remove ………236
（振動・雑音などを）除去する suppress ………268
（物・事を）除去する eliminate ………123
（物を）所有する have ………159
（～を）処理する deal ………101
（事を）処理する handle ………158
（信号・データなどを）処理する process ………219
（物・事を）処理する treat ………280
（物・事を）調べる check ………74
（可能性などを）調べる explore ………138
（物・事を）調べる investigate ………175
（物・事を）調べる study ………263
（計画などを）遂行する execute ………134
（能力・特性などを）推定する estimate ………130

（事を）推定する　presume ……217
（物・事を）図示する　illustrate ……165
（実験・解析などを）する　do ……120
（物・事を）～にする　make ……186
（仕事・試験などを）する　perform ……210
（注意・決心・見方などを）する　take ……272
（～の）寸法をしている　measure ……191
（物・事を）制御する　control ……95
（事を）制限する　restrict ……242
（物・事を）制限する　limit ……182
（回路などを）整合する　match ……189
（装置・機械・製品などを）製作する　fabricate ……143
（物を）製作する　make ……186
（薄膜・結晶などを）成長させる　grow ……156
（物・事が）成長する　grow ……156
（物・事が）～のせいにする　attribute ……59
（回路などを）絶縁する　isolate ……177
（数量・状態・場所などに）接近する　approach ……55
（物・事が）接近する　approach ……55
（物を）設計する　design ……110
（二つ以上の物が）接続する　connect ……89
（二つ以上の物を）接続する　connect ……89
（物を）設置する　place ……212
（物・事を）設定する　set ……254
（物・事を）セットする　set ……254
（複数個の中から～を）選択する　choose ……75
（物・事を）選択する　select ……249
（物・事を）選択する　pick ……211
（大きさ・数量・程度などを）増加させる　increase ……169
増加する　build up ……66
（物と）相互作用する　interact ……174
（物・事を）相殺する　cancel ……70
（機械・装置などを）操作する　operate ……206
（信号・情報などを）送信する　send ……251

（電波・信号・データなどを）送信する transmit ………278
増大する build up ………66
（価値・性能などを）増大する enhance ………126
（物・事が）増大する grow ………156
（大きさ・数量・程度などが）増大する increase ………169
（物・事が）相当する correspond ………99
（物を）装備する equip ………129
（電気量・信号などを）増幅する amplify ………50
（反応などを）促進する accelerate ………41
（事を）促進する facilitate ………144
（位置・形状などを）測定する determine ………112
（長さ・量・大きさなどを）測定する measure ………191
（物を）備えつける equip ………129
（物を）備える provide ………225
（物が）存在する exist ………136
（物・事が）存在する occur ………204

た行

（物・事が）対応する react ………230
（薄膜などを）堆積させる deposit ………107
（価値・性能などを）高める enhance ………126
（性能・特性・価値などを）高める improve ………167
（物・事を）確かめる check ………74
（証拠・予測などを）確かめる confirm ………88
（事を）確かめる verify ………288
（人・物事を）助ける assist ………57
（人・物事が）助ける assist ………57
（物・事を）助ける help ………161
（物・事の）由来をたずねる derive ………108
（物が）達する come ………80
（ある状態・結果などに）達する reach ………229
（仕事などを）達成する accomplish ………42
（仕事・目的などを）達成する achieve ………43
（目的・望みなどを）達成する attain ………59

（建築・構造物などを）建てる build ……66
（信号・データなどが）たまる collect ……79
（物・事を）試す try ……281
（物・事をある状態のまま）保つ hold ……162
（ある動作・状態を）保つ keep ……178
（物・事を）単純化する simplify ……256
（数量・状態・場所などに）近づく approach ……55
（物・事が）近づく approach ……55
（データを）蓄積する store ……261
（物・事に）注意する note ……203
（信号・情報・データなどを）抽出する extract ……142
（物・事に）注目する note ……203
（物・事を）調査する examine ……133
（可能性などを）調査する explore ……138
（物・事を）調査する investigate ……175
（機械・装置・電気量などを）調整する adjust ……46
（ある値に）調整する tune ……282
（機械・装置・電気量などを）調節する adjust ……46
（エネルギーなどを）貯蔵する store ……261
（時間を）費やす take ……272
（物を）通過させる pass ……209
（物が）通過する pass ……209
（人・物が）通信する communicate ……81
（物・事が）通じる lead ……181
（物・事を）使う use ……284
（物が）着く come ……80
（電気・熱・信号などを）作り出す generate ……154
（物・事を）作り出す produce ……220
（装置などを）作る build ……66
（物を）作る form ……151
（物を）作る make ……186
（他の物に）つけ足す add ……45
（音・光・電磁波・熱などを）伝える propagate ……221
（光・熱などを）伝える transmit ……278

（音・光・電磁波・熱などが）伝わる　propagate ……221
（光・音などが）伝わる　travel ……279
（ある動作・状態を）続ける　keep ……178
（物・事が）～し続ける　keep ……178
（二つ以上の物が）つながる　connect ……89
（二つ以上の物を）つなぐ　connect ……89
（物を）詰め込む　fill ……147
（事を）提案する　propose ……222
（計画・考えなどを）提案する　suggest ……265
（性能・質などを）低下させる　degrade ……104
（速度・温度・程度などを）低下させる　lower ……184
（語などを）定義する　define ……103
（物・事を）提供する　offer ……205
（数量・程度・範囲などを）低減させる　reduce ……233
（振動・雑音などを）低減する　suppress ……268
（事を）定式化する　formulate ……152
（論文・報告書などを）提出する　present ……217
（誤りなどを）訂正する　correct ……98
（状況などに）適応する　adapt ……44
（状況などに）適合させる　adapt ……44
（物・事と）適合する　match ……189
（物・事に）適合する　suit ……266
（物・事に）適する　suit ……266
（方法・規則などを）適用する　apply ……53
（物・事を）点検する　check ……74
（データなどを）転送する　transfer ……276
（信号などを）ディジタル化する　digitize ……115
～できない　fail to do ……145
（電波・信号・データなどを）伝送する　transmit ……278
（情報・データなどを）伝達する　communicate ……81
（音・光・電磁波・熱などが）伝播する　propagate ……221
（音・光・電磁波・熱などを）伝播させる　propagate ……221
（音・光・電磁波・熱などを）伝搬させる　propagate ……221
（音・光・電磁波・熱などが）伝搬する　propagate ……221

（回路・信号などを）同期する　synchronize ……270
（物を）統合する　integrate ……172
（機械・装置などを）動作させる　operate ……206
（〜に対して）動作する　behave ……64
（機械・装置などが）動作する　operate ……206
（機械などが）動作する　perform ……210
（プログラム・機械・装置などが）動作する　run ……245
（機械などが）動作する　work ……290
（物を）搭載する　mount ……197
（回路・信号などを）同調する　synchronize ……270
（受信機・レーザなどを）同調させる　tune ……282
（物・事を）同定する　identify ……163
（物を）通す　pass ……209
（物が）通る　pass ……209
（ケーブル・配線などが）通る　run ……245
（問題・数式などを）解く　solve ……258
（物・事を）特徴づける　characterize ……73
（物・事に）取って代わる　replace ……238
（物・事が）とどまる　remain ……236
（塗料などを）塗布する　coat ……78
（物・事を）伴う　involve ……177
（仕事・問題などに）取り組む　address ……46
（問題・研究などに）取り組む　approach ……55
（物・事を）取り消す　cancel ……70
（物を）取り付ける　mount ……197
（不要なものを）取り除く　remove ……236
（データなどを）とる　take ……272
（形状・性質などを）とる　take ……272

な行

（物・事を）内蔵する　include ……168
（電流・液体などが）流れる　flow ……150
（仕事などを）成し遂げる　accomplish ……42
（仕事・目的などを）成し遂げる　achieve ……43

（目的・望みなどを）**成し遂げる** attain ……59
（名前などを）**名づける** call ……69
（式などが）**成り立つ** hold ……162
（物・事が）**〜になる** become ……62
（部分・要素から）**なる** consist ……91
（塗料などを）**塗る** coat ……78
（物を）**熱する** heat ……160
（物・事を）**除く** eliminate ……123
（物・事について）**述べる** describe ……109
（事実・意見・問題などについて）**述べる** state ……261
（物・事が）**〜のままでいる** remain ……236

は行

（物を）**配置する** place ……212
（事が）**始まる** begin ……63
（事が）**始まる** start ……260
（事を）**始める** begin ……63
（事を）**始める** start ……260
（〜の役割を）**果す** play ……213
（機械などが）**働かなくなる** fail ……145
（機械などが）**働く** act ……44
（物・事を）**発見する** find ……148
（問題・困難などが）**発生する** arise ……56
（物・事を）**発生させる** produce ……220
（電気・熱・信号などを）**発生させる** generate ……154
（物・事が）**発展する** grow ……156
（熱・光・音・ガスなどを）**放つ** emit ……124
（光・音・熱などを）**反射する** reflect ……234
（物・事が）**反応する** react ……230
（事が）**判明する** prove ……223
（物・事と）**比較する** compare ……82
（結果として〜を）**引き起こす** cause ……71
（物・事を）**引き起こす** produce ……220
（性質・知識・利益などを）**引き出す** derive ……108

（線などを）引く　draw……121
（物体・信号・波形を）歪ませる　distort……118
（物・事を）必要とする　need……200
（物・事を）必要とする　require……241
（数量・大きさなどの点で）等しい　equal……128
（塗料などを）被覆する　coat……78
（物を）冷やす　cool……98
（価値・性質などを）評価する　assess……56
（能力・特性などを）評価する　estimate……130
（価値・能力などを）評価する　evaluate……131
（物・事を）表現する　represent……240
（事を）表現する　express……140
（文字・図形などを）表示する　display……117
（信号などを）標本化する　sample……246
（大きさ・数量・程度などが）増える　increase……169
（変調波を）復調する　demodulate……105
（物を）含む　contain……94
（物・事を）含む　include……168
（物を）含む　incorporate……168
（物・事を）含む　involve……177
（物・事を）防ぐ　prevent……218
（大きさ・数量・程度などを）増やす　increase……169
（～に対して）ふるまう　behave……64
（図形などを）プロットする　plot……213
（物・事を）分割する　divide……119
（物・事を）分析する　analyze……51
（物を）分離する　isolate……177
（物を）分離する　separate……253
（物・事を）分類する　classify……77
（数量などを）平均して～となる　average……60
（数量などを）平均する　average……60
（数・量などを）減らす　decrease……102
（数量・程度・範囲などを）減らす　reduce……233
（数・量などが）減る　decrease……102

（物・事を）変化させる　vary……287
（物・事が）変化する　change……72
（物・事が）変化する　vary……287
（信号などを）変換する　convert……96
（エネルギー・信号などを）変換する　transduce……276
（物を）変換する　trasform……278
（物体・信号・波形を）変形させる　distort……118
（物を）変形する　trasform……278
（光を）偏光させる　polarize……214
（物・事を）変更する　change……72
（部分的に）変更する　modify……194
（周波数・振幅などを）変調する　modulate……195
（物・事を）妨害する　interfere……175
（研究・調査などを）報告する　report……239
（熱・光・音・ガスなどを）放射する　emit……124
（電磁波・光・熱などを）放射する　radiate……228
（熱・光・音・ガスなどを）放出する　emit……124
（データを）保管する　store……261
（損傷・危険などから）保護する　protect……223
（物を）保持する　hold……162
（状態・温度・圧力・性能などを）保持する　maintain……185
（損失などを）補償する　compensate……83
（事を）保証する　ensure……127
（物・事を）保証する　guarantee……157
（装置・回路などを）補正する　correct……98
（データ・情報などを）保存（記憶、格納）する　hold……162
（物・事を）施す　give……155

ま行

（物を）まとめる　integrate……172
（損傷・危険などから）守る　protect……223
（要件・条件などを）満足させる　satisfy……247
（物・事を）見いだす　find……148
（物を）満たす　fill……147

(要件・条件などを) 満たす satisfy ……247
(性質・知識・利益などを) 導き出す derive ……108
(結論などを) 導く draw ……121
(物・事を) 見つける find ……148
(物・事を) 認める observe ……203
(～だと) 見なす assume ……58
(物を) 見る view ……289
(物・事の間の相違を) 見分ける distinguish ……118
(物・事がある結果・状態に) 向かう tend ……274
(物・事を) 無視する ignore ……164
(物・事を) 無視する neglect ……201
(考え方・理論などを) 明確にする clarify ……76
(事を) もたらす bring ……65
(結果として～を) もたらす cause ……71
(物・事を) もたらす yield ……293
(物・手段などを) 用いる employ ……125
(物を) もっている have ……159
(物を) 持ってくる bring ……65
(物・事を) モデル化する model ……193
(数式などの) 値を求める evaluate ……131
(物・事を) 求める find ……148

や行

(金属などを) 焼きなます anneal ……52
(物・事を) 役立たせる utilize ……286
(性質・特性などを) 有する have ……159
(電流・磁気などを) 誘導する induce ……171
～を許す allow ……49
(物・事を) 用意する prepare ……216
(事を) 容易にする facilitate ……144
(物・事を) 要求する require ……241
～のようだ appear to do ……52
(物・事を) 要約する summarize ……266
(物・事を) 予期する expect ……136

（振動・雑音などを）抑制する suppress ……268
（物・事を）予想する expect ……136
（事を）予測する predict ……215
（名前などを）呼ぶ call ……69
（人が）呼ぶ call ……69

ら行

（事を）理解する understand ……283
（証拠・予測などを）立証する confirm ……88
（理論・学説などを）立証する establish ……129
（事を）立証する prove ……223
（事を）立証する verify ……288
（信号・エネルギーなどを）量子化する quantize ……227
（物・手段などを）利用する employ ……125
（特性などを）利用する exploit ……137
（物・事を）利用する use ……284
（物・事を）利用する utilize ……286
（物を）冷却する cool ……98
（電磁石・コイルなどを）励磁する excite ……134
（物・事を）例示する illustrate ……165
（アンテナなどを）励振する excite ……134
（性能・質などを）劣化させる degrade ……108
（物・事を）列挙する list ……184
（二つ以上の物が）連結する connect ……89
（二つ以上の物を）連結する connect ……89
（物を）連結する couple ……100
（信号などを）ろ波する filter ……148
（学説・真理などを）論証する demonstrate ……105
（～を）論じる deal ……101
（物・事について）論じる discuss ……116
（物・事を）論じる treat ……280

わ行

（物・事を）わかっている know ……179

（事が）わかる　prove ……223
（事が）わかる　see ……249
（物・事を）分ける　divide ……119
（物を）分ける　separate ……253
（範囲・程度などが）わたる　range ……229

A

absorb

(absorbed, absorbed, absorbing)

〈他〉（液体・音・光などを）吸収する

☐ Light is incident from the bottom and is absorbed in the InGaAs absorption layer of thickness l_a.

（光は底部から入射し,そして厚さがl_aのInGaAs吸収層中に吸収される）

☐ It is found that less than 1％ of the power of the incident pulse can be absorbed by the substrate material.

（入射パルスのパワーの１％未満のパワーが基板材料に吸収される可能性がある，ということがわかる）

☐ The system does not have the capacity to absorb the excess energy.

（そのシステムには，余分のエネルギーを吸収する容量はない）

☐ The absorbed light causes a decrease of the electrode voltages according to the following relation:

（吸収光は次の関係の通りに，電極電圧の低下を引き起こす）

ポイント☞n:absorption 吸収　　adj:absorbable 吸収性の

☞absorber は吸収器・吸収体.

accelerate

(accelerated, accelerated, accelerating)

〈他〉（物質などを）加速する

（反応などを）促進する

☐ The crosslinkng process can be accelerated by heating the material.

（架橋過程は材料を加熱することで促進できる）

☐ The computation of the current coefficients is accelerated by using FFT algorithm.
（電流係数の計算は，FFTアルゴリズムを用いることで速くなる）

☐ The average electron energy corresponding to the electrons having been accelerated in electric field up to E_c = 3.5 kV/m can be estimated from modeling using nonstationary approaches.
（E_c = 3.5 kV/mまでの電場で加速された電子に対応する平均電子エネルギーは，非定常アプローチを用いたモデリングから推定できる）

☐ The development of high quality material will accelerate the progress of the potential superconductive applications.
（高品質材料の開発によって，超伝導の可能な用途の進展は促進するだろう）

☐ This accelerates the speed of response of the system.
（このことで，システムの応答速度は速まる）

ポイント☞［反］decelerate(減速する)　　n:acceleration 加速
☞acceleratorは加速器．particle accelerator(粒子加速器)

accomplish

(accomplished, accomplished, accomplishing)

〈他〉（仕事などを）成し遂げる，達成する

☐ Scanning of the antenna beam is accomplished by moving the feed along the perimeter of the lens surface.
（アンテナビームの走査は，レンズ面の周辺に沿ってフィードを動かすことで成し遂げられる）

☐ We accomplished this goal through a comprehensive study of the transmission line propagation factor.
（伝送線路の伝搬係数の包括的な研究によって，この目標を達成した）

ポイント☞［類］achieve, attain　n:accomplishment達成　adj:accomplished 達成した

☞「ある仕事・目的を努力してうまくやり遂げる」のニュアンスがある.

achieve

(achieved, achieved, achieving)

〈他〉（仕事・目的などを)成し遂げる，達成する

☐ The calibration of the network analyzer can be achieved in two stages.

(このネットワークアナライザの校正は，二段階で達成できる)

☐ These temperatures can only be achieved by employing expensive liquid helium refrigeration units.

(これらの温度は高価な液体ヘリウム冷却装置を利用することでのみ,達成できる)

☐ Recently,rapid progress in GaAs metal-semiconductor-metal (MSM) photodetector performance has been achieved.

(GaAs 金属－半導体－金属(MSM)光検出器の性能の急速な進歩が,最近になって成し遂げられた)

☐ This IC achieves a bandwidth of 3.5 GHz,and a high gain of 30 dB.

(このICは3.5 GHzの帯域幅，30 dBの高利得を達成する)

☐ The floating-point data format is necessary to achieve the high accuracy for position measurement.

(位置測定に対して高精度を達成するためには,浮動小数点データ形式が必要である)

☐ Vector quantization schemes offer the potential of achieving high-quality speech at 9.6 kbit/s.

(ベクトル量子化方式は,9.6 kbit/sにおいて高品質音声を達成する可能性をもたらす)

ポイント☞[類] accomplish, attain　　n:achievement 達成

☞「能力・技術を使って困難を乗り越えて目標を達成する」のニュアンスがある．

act

(acted,acted,acting)

〈自〉 (機械などが)作動する，働く
(力などが)作用する
act as～：～の役目を果す
act on[upon]～：～に作用する

☐ The buses act to partition the chip into functional modules.
(これらのバスは，チップを機能モジュールに分割するように働く)

☐ The amplitude modulation is performed by an analog multiplier that acts as a balanced modulator.
(その振幅変調は，平衡変調器の役目をするアナログ乗算器によって行われる)

☐ The forces acting on the vehicle are the traction force F_t and the lateral forces F_f,　F_r.
(車に作用する力は，牽引力F_tと横方向の力F_f, F_rである)

ポイント☞名詞としては「行為・行動」. 同じ意味の名詞には action もある．

☞in the act of doing(～している最中に)

adapt

(adapted, adapted, adapting)

〈他〉 (状況などに)適合させる，適応させる，合わせる
adapt A to B：AをBに適応させる，応用する

☐ The control logic can easily be adapted to other single-board computers.
(その制御論理は他のシングルボードコンピュータに容易に適合できる)

☐ This finite element method is well adapted to the static analysis of local

properties of a recording head.
（この有限要素法は，記録ヘッドの局所的性質の静的解析にうまく適合する）

☐ We adapt this second approach to fuzzy systems in this paper.
（本論文では，この第二の方法をファジィシステムに応用する）

ポイント ☞n:adaptation 適合・適応　　adj:adaptable 適応できる
☞adaptor(adapter)はアダプター.

add

(added, added, adding)

〈他〉（他の物に）加える，つけ足す
add A to B：AをBに加える

☐ Usually, when an increase of this ratio is needed, an analog noise is added to the input signal.
（普通，この比を大きくする必要がある場合，アナログ雑音を入力信号に加える）

☐ The converted current is added directly to the output current of MP.
（変換された電流をMPの出力電流に直接加える）

☐ In practice a random noise is added, before conversion, to the measured signal.
（実際には，変換前にランダムノイズは測定信号に加えられる）

☐ We added noise to the synthesized signal in Experiment 1.
（実験1において，雑音を合成信号に加えた）

☐ Added to Fig.3 is the doping concentration along the Si/SiO_2 interface.
（図3に加えたのは，Si/SiO_2界面に沿ったドーピング濃度である）

☐ The large J is obtained by adding extra inertia to the rotor.
（大きなJは，余分の慣性を回転子に加えることで得られる）

ポイント☞n:addition 追加・加算　adj:additional 追加の・余分の

☞引き算はsubtraction, 掛け算は multiplication, 割り算は division である.

☞in addition(さらに, その上)

address

(addressed, addressed, addressing)

〈他〉（仕事・問題などに）取り組む

（問題などを）扱う

□ These are complicated many-body problems that must be addressed in any model of high-temperature superconductors.
（高温超伝導体のどんなモデルでも, 取り組まなければならない複雑な多体問題が存在する）

□ Having solved the inverse problem, we can now address this problem.
（逆問題を解いたので, 今ではこの問題を扱うことができる）

□ This paper addresses the robust stability problem for time-delay systems with parametric uncertainties.
（本論文では, パラメトリック不確実性をもつ時間遅れシステムに対するロバスト安定性問題を取り扱う）

ポイント☞名詞としては「住所・アドレス」. コンピュータ用語では「アドレス指定する」の意味で使われ, addressable はアドレス指定可能な.

adjust

(adjusted, adjusted, adjusting)

〈他〉（機械・装置・電気量などを）調節する, 調整する

□ The output voltage can be adjusted precisely by changing the resistance R_l.
（出力電圧は, 抵抗R_lを変えることで精密に調整できる）

□ The frequency shift can easily be adjusted by conventional methods.

(その周波数偏移は従来の方法で容易に調整できる)

☐ This robot hand control system automatically adjusts the grip to suit the shape and weight of an object.

(このロボットハンド・制御システムは,物体の形状と重量に適合するようにグリップを自動調節する)

☐ The gain correction block adjusts the current gain according to the predicted signal value.

(その利得補正ブロックは,予測信号値によって電流利得を調整する)

ポイント☞[類] regulate　n:adjustment 調節・調整　adj:adjustable 調節(調整)可能な

☞「正しい,あるいは望ましい状態にもっていく」のニュアンスがある.

---

adopt

(adopted, adopted, adopting)

〈他〉(理論・技術・方式などを)採用する,使用する

☐ Recently,modern control techniques have been adopted in HDD servo design to significantly improve control performance.

(最近,制御性能を著しく向上させるために,HDDサーボの設計に現代制御技術が採用された)

☐ We adopt a Gaussian distribution of potentials for convenience but the theory we develope in this paper can be applied to any other potential distribution.

(便宜上,ガウス電位分布を採用するが,本論文で開発する理論は,他の電位分布にも適用できる)

☐ To overcome this problem we adopt the following procedure,in which the range of integration is divided into two parts $0 \leqq x \leqq 1/b$ and $1/b < x < \infty$.

(この問題を解決するために,積分範囲が $0 \leqq x \leqq 1/b$ と $1/b < x < \infty$ の二つの部分に分けられる,以下の手順を採用する)

ポイント☞n:adoption 採用

☞新しい技術・理論・方針などを採用する場合に用いられる.

affect

(affected, affected, affecting)

〈他〉(物・事に)影響を及ぼす

☐ The scale factor is affected by thermal expansion of the coil.
(目盛係数はコイルの熱膨張の影響を受ける)

☐ The speed performance of a CMOS gate is strongly affected by the initial delay times.
(CMOSゲートの速度性能は，初期遅延時間の影響を強く受ける)

☐ This can seriously affect the performance of the current sensors.
(これは電流センサの性能に大きな影響を及ぼしかねない)

☐ A magnetic field up to at least 1 tesla does not affect the phosphor calibration.
(少なくとも1テスラまで，磁場は蛍光体の校正に影響を及ぼさない)

☐ The induced magnetic anisotropy affects the domain structure and hence the magnetic properties,especially the permeability.
(その誘導磁気異方性は磁区構造に影響を及ぼし，それゆえに磁気特性，とりわけ透磁率に影響を及ぼす)

ポイント☞［類］influence

☞affect は「直接的」に影響を及ぼす場合に，influence は「間接的」に影響を及ぼす場合に用いられるとされるが，技術英語では両者に差異はほとんどない．名詞としては「感情」であり，「影響」という意味はない．

agree

(agreed, agreed, agreeing)

〈自〉（物・事が）一致する

agree with A：Aと一致する

□ This result qualitatively <u>agrees</u> with the resluts described in Section Ⅱ.
（この結果は第2節で述べられた結果と定性的に一致する）

□ This value <u>agrees</u> well with that of 1.1×10 ^{5} m/s estimated from f_T measurement.
（この値は，f_Tの測定から推定された1.1×10^5 m/sの値によく一致している）

□ It can be seen that the theoretical results <u>agree</u> very well with the experimental results.
（理論から得られた結果は実験結果と非常によく一致していることが理解できる）

ポイント☞［反］disagree(一致しない)　　［類］coincide, accord, conform, concur
n:agreement 一致
☞agree with＝be in agreement with.「よく一致する」は agree well with である.

allow

(allowed, allowed, allowing)

〈他〉～を許す，～を許容する，～を可能にする

allowA to do：Aが～するのを認める

〈自〉allow for～：～を考慮に入れる，～を可能にする

□ Different equalizers will be <u>allowed</u> in this GSM system.
（このGSMシステムでは，さまざまな等化器が使用できる）

□ A slip surface is <u>allowed</u> to exist in the airgap between stator and rotor.
（固定子と回転子の間のエアギャップ内には，すべり面が存在することができる）

☐ The design of the program allows the students to perform dynamic security analysis of a power system.
(そのプログラムを設計することで,学生は電力系統の動的安全解析を行えるようになる)

☐ Measurements of rare-earth ion magnetic resonance allow us to study effectively this interaction.
(希土類イオン磁気共鳴の測定値によって,この相互作用を効果的に研究できる)

☐ Fieldbus allows for collection and management of massive amounts of process data.
(フィールドバスはばくだいな量のプロセスデータを収集・管理できるようになっている)

☐ The approach used here simultaneously measures the optical phase change of the light beam and the complex reflection coefficient of the material, thus allowing phase compensation to be applied.
(ここで用いた手法は,光線の光学位相変化と材料の複素反射係数を同時に測定するので,位相補償が適用できる)

ポイント ☞［反］forbid(禁ずる)
☞「妨げたりしないで自由にさせておく」というニュアンスがある.

amplify
(amplified, amplified, amplifying)
〈他〉（電気量・信号などを）増幅する

☐ At FM front-end of a radio a high frequency signal about 100 MHz is amplified.
(ラジオのFMフロントエンドで，約100 MHzの高周波信号は増幅される)

☐ The output voltage of the thermopiles is amplified by the differential amplifier.
(サーモパイルの出力電圧は差動増幅器によって増幅される)

□ The differential voltage across the detector resistances is amplified by a factor of 70.
(検出器の抵抗器両端の差動電圧は，70倍増幅される)

□ This transistor can amplify the photocurrent.
(このトランジスタは光電流を増幅できる)

□ Amlified noise is detected at the integrator output.
(増幅された雑音は積分器の出力側で検出される)

ポイント☞n:amplification 増幅
☞amplifier は増幅器． power amplifier(電力増幅器)， operational amplifier(演算増幅器)， optical amplifier(光増幅器)

analyze
(analyzed, analyzed, analyzing)
〈他〉（物・事を）解析する，分析する

□ The ladder networks can be analysed very easily and accurately.
(そのはしご型回路網は非常に容易に，そして正確に解析できる)

□ The waveguide can be analyzed by a large number of techniques reported in the literature.
(その導波管は文献で報告された多くの手法によって解析できる)

□ The sample used for this examination was obtained from the same wafer which was analysed by X-rays.
(この試験で用いられた試料は, X線で分析されたものと同一のウェハから得られた)

□ We analyze the CPFSK performance when the receiver is subject to narrowband Gaussian noise interference.
(受信機が狭帯域ガウス雑音干渉を受ける場合のCPFSK性能を分析する)

ポイント☞analyse も使われる.
☞n:analysis(複数形 analyses) 解析・分析　adj:analytical 分析(解析)的な
☞analyzer は解析器・アナライザ，analyst は分析者・アナリスト.

anneal
(annealed, annealed, annealing)
〈他〉（金属などを)焼きなます，アニールする

□ The wafer is annealed at 810℃ for 20 min.
(ウェハを20分間，810℃でアニールする)

□ The samples were annealed in an oxygen atmosphere at 900℃ for 24 hours and then slowly cooled to room temperature.
(試料を酸素雰囲気中において，24時間にわたって，900℃で焼きなましし，それから室温までゆっくりと冷ました)

□ One way of doing this is to anneal the material in a magnetic field.
(これを行う一つの方法は，磁場中で材料をアニールすることである)

□ For these samples no oxide layer was formed when annealed under the same condition as above.
(これらの試料の場合，上と同じ条件でアニールされる場合には,酸化層は形成されなかった)

ポイント☞n:annealing 焼きなまし・アニール
☞半導体では「アニールする」という.

appear
(appeared, appeared, appearing)
〈自〉（物・事が)現れる，生ずる，～のように見える
appear［to be］A：Aのように思われる(見える)
appear to do～：～のようだ

□ The first papers on integrated scanners on glass substrates appeared in 1984.

(ガラス基板上の集積スキャナに関する初めての論文は，1984年に現れた)

☐ The waveform appears as a burst of noise with a duration of about 30 μs.
(その波形は持続時間が約30μsの雑音のバーストとして現れる)

☐ The method appears to be accurate and computationally efficient.
(その方法は正確で，計算効率がいいように思われる)

☐ This problem appears to present serious difficulties and is presently unsolved.
(この問題は，深刻な困難を引き起こすらしく，現在未解決である)

☐ The convolutions appearing in (4) and (5) can be efficiently computed by using the FFT.
((4)と(5)に見られる畳込みは，FETを用いることで効率良く計算できる)

ポイント☞［反］disappear(消える)　［類］emerge　n:appearance 出現
☞“appear [to be] A”のAは，名詞・形容詞・前置詞句．

apply
(applied, applied, applying)
〈他〉（方法・規則などを）適用する，応用する，使用する
（力・熱・電圧などを）加える，印加する
apply A to B：AをBに適用する，加える
〈自〉（方法・規則などが）あてはまる
apply to A：Aにあてはまる

☐ However,this method cannot be applied easily because R_s is not known.
(しかしながら，R_sが未知なので，この方法は容易に適用できない)

☐ Two-stage template matching can in principle be applied in any application where template matching is appropriate.
(二段階テンプレートマッチングは，テンプレートマッチングが妥当な応用に対して原則として適用できる)

□ Voltage was <u>applied</u> on these electrodes, that is, across the piezoelectric polymer.
（電圧はこれらの電極上，すなわち圧電ポリマーの両端に印加された）

□ This spectrometer was <u>applied</u> to study magnetic resonance in Nd_2CuO_4.
（この分光計はNd_2CuO_4における磁気共鳴を研究するのに用いられる）

□ This transform can be <u>applied</u> to double integrals with singularities.
（この変換は特異点をもつ二重積分に適用できる）

□ Image compression is often <u>applied</u> to fields such as television broadcasting and image storage.
（画像圧縮はテレビ放送や画像記憶といった分野でしばしば応用される）

□ In this paper we shall <u>apply</u> the multiple Fourier series to estimate the d-dimensional probability density and discriminant functions.
（本論文では，d次元確率密度関数と判別関数の推定に多重フーリエ級数を適用する）

□ Therefore, we can easily <u>apply</u> these impedances to parameter conversions for microwave amplifier design.
（したがって，マイクロ波増幅器の設計のために，これらのインピーダンスをパラメータ変換に容易に適用できる）

□ <u>Applying</u> this equation,we obtain the dielectric constant data for eight different samples, as shown in Table Ⅱ.
（この方程式を適用すれば，表Ⅱに示されるように，8個の異なる試料に対する誘電率のデータが得られる）

ポイント ☞n.application 適用・応用　　adj.applicable 適用できる
☞“apply to A”では進行形は不可．

approach

(approached, approached, approaching)

〈他〉（数量・状態・場所などに）近づく，接近する
（問題・研究などに）取り組む
〈自〉（物・事が）近づく，接近する

☐ When α approaches 180°, the diffracted field becomes weaker.
（αが180°に近づく場合は，回折場は弱くなる）

☐ For more complicated structures, errors in estimating σ approached 50% even if the error in estimating the mean was within 5%.
（より複雑な構造物の場合，たとえ平均値を推定する時の誤差が5%以内にあっても，σを推定する時の誤差は50%に近づく）

☐ This situation is not physically accurate because the breakdown voltage must asymptotically approach unity as the normalized radius of curvature is increased.
（正規化された曲率半径が長くなるにつれて，破壊電圧は漸近的に1に近づくため，この状態は物理的に正確ではない）

☐ As the carrier frequency approaches to the resonant frequency, the number of the resonant pulses in one-cycle of the carrier decreases.
（搬送周波数が共振周波数に近づくにつれて，1サイクルの搬送波に含まれる共振パルス数は減少する）

☐ In time-domain frequency stability measurements, one should try to approach the optimum conditions to minimize the equvalent bandwidth.
（時間領域周波数安定性を測定する際，等価帯域幅を最小にするための最適条件に取り組むべきである）

ポイント☞名詞としては「接近・アプローチ」．他動詞として用いる場合には，to は不要なので注意．

arise

(arose, arisen, arising)

〈自〉（問題・困難などが）生じる，起こる，発生する

arise from A：Aから生じる，起こる，発生する

☐ For a general defect a problem arises in this treatment.
（一般的な欠陥の場合，この処理には問題が生じる）

☐ No magnetic moments arise from orbital motion in ferromagnetic substances.
（磁気モーメントは強磁性体における軌道運動からは生じない）

☐ The temperature-dependence in the I/V characteristic arises principally from the parameters I_s and V_T of the characteristic equation.
（I／V特性の温度依存性は，主として特性方程式のパラメータであるI_sとV_Tに由来する）

☐ Two problems arising in lossy LC ladder network synthesis have been treated.
（損失性LCはしご型回路網の合成で生じる二つの問題を取り扱った）

ポイント☞［類］happen, occur, take place, give rise to, result from

assess

(assessed, assessed, assessing)

〈他〉（価値・性質などを）評価する，算定する，決定する

☐ The performance of the designed control system can be assessed from the responses depicted in Fig.5.
（設計された制御システムの性能は，図５に描かれた応答より評価できる）

☐ The results were used to assess the effects of third-harmonic fluxes in rotating machines.
（回転機械の第三高調波磁束の影響を決定するために，これらの結果を用いた）

☐ An estimate of the average power spectrum of the noise is useful in assessing the probability of flaw detection.
(雑音の平均パワースペクトルの推定値は，傷の検出確率を算定する際に役立つ)

ポイント☞［類］evaluate, estimate, value,appreciate　　n:assessment 評価
☞「性能・重要性といった物の価値を査定する」のニュアンスがある．

assist
(assisted, assisted, assisting)

〈他〉（人・物事を）助ける，支援する，援助する
assist A in doing：Aが～するのを助ける
〈自〉（人・物事が）助ける，支援する，援助する
assist in A：Aを助ける

☐ The control algorithm is written in the assembly language for the DSP and a software development is assisted by a host personal computer NEC PC-9801 under the MS-DOS environment.
(DSPのために制御アルゴリズムはアセンブリ言語で書かれ，ソフトウェア開発はMS－DOS環境のもとで，ホストパーソナルコンピュータNEC PC-9801によって支援される)

☐ A thorough understanding of ultrasonic waveguides will assist the design of these biopsy needles and catheters.
(超音波導波管を完全に理解することは，これらの生検針とカテーテルの設計の助けとなるだろう)

☐ We hope to assist blind people in making an internal model of their surroundings using this chip.
(盲人がこのチップを用いて，自分達の環境の内部モデルを構築するのを支援したいと思う)

☐ Expert systems can assist in operator training.

（エキスパートシステムはオペレータの訓練の助けとなる）

☐ Design verification procedures, assisted by computer-aided design (CAD) tools, are used to eliminate the design defects.
（計算機援用設計（CAD）ツールによって支援された設計検証手順は，設計時の欠陥を除去するのに用いられる）

ポイント☞［類］support, help　　n:assistance 援助
☞「補助的に助ける」のニュアンスがある．

assume

(assumed, assumed, assuming)

〈他〉（～だと）仮定する，考える，見なす
assume A to be B：AをBと仮定する［考える，見なす］
assuming(assume)that節：～と仮定して

☐ Throughout this paper, a synchronous clock with a period of 100 ns is assumed.
（本論文を通じて，周期が100 nsの同期クロックを仮定する）

☐ We assume that an electric field E is incident upon an inhomogeneous dielectric body.
（電場Eは不均一誘電体に入射するものと仮定する）

☐ We assume that these variations are so slow that the phase jitter can be modeled as a random variable with known probability density function $f(\Theta)$.
（これらの変動は非常に遅いので，位相ジッターは確率密度関数$f(\Theta)$が既知の確率変数としてモデル化できるものと仮定する）

☐ We assumed the heat uniformly generated inside the active region to be 20 mW.
（活性領域内部から均一に発生する熱は，20 mWだと見なした）

☐ The surface S of the waveguide is assumed to be a perfect electrical conductor.
（導波管の表面 Sは完全導体だと仮定する）

□ Assuming that H_{x1} and H_{x2} in Eq. (18) equal to those in Eq. (17), respectively, the following equation can be obtained as a sum of Eqs. (17) and (18).
（方程式(18)中のH_{x1}とH_{x2}は，それぞれが方程式(17)のと等しいと仮定すれば，次の方程式は方程式(17)と(18)の和として得られる）

□ The radiation pattern for the diagonally fed square horn was calculated by assuming that orthogonal TE_{10} modes are excited with equal amplitudes.
（対角線方向に給電された正方形ホーンの放射パターンは，直交TE_{10}モードが同じ振幅で励振されると仮定することで計算された）

ポイント☞［類］suppose, presume, hypothesize　　n:assumption 仮定

attain

(attained, attained, attaining)

〈他〉（目的・望みなどを）達成する，成し遂げる，得る

□ The maximum of the effective coupling is attained for natural modes that have the same frequencies in the uncoupled state.
（実効的な結合の最大値を，非結合状態のと同じ振動数をもつ固有モードに対して得る）

□ A superscalar RISC processor which attains high operation speed is discussed.
（高演算速度を達成するスーパースカラーRISCプロセッサについて論じる）

ポイント☞[類] accomplish, achieve　　n:attainment 達成

☞「達成困難な目標を努力の末，成し遂げる」のニュアンスがある．

attribute

(attributed, attributed, attributing)

〈他〉（物・事が）～にあると考える，～に帰する，～のせいにする
attribute A to B：AがBにあると考える，AをBに帰する

□ The errors are mainly attributed to the environmental noise during the measurement.

(その誤差は，主に測定中の環境騒音による)

☐ The variation can be mainly attributed to the Fabry-Perot type interference between reflectors in the system, because the frequency of light and the length of fiber change by the temperature fluctuation.
(その変化は，主にシステム内の反射器間のファブリーペロ型干渉によるものと考えられる。なぜなら，光の周波数とファイバの長さは温度の変動によって変化するからである)

☐ This effect can be attributed to the series resistance of the device structure.
(この効果は素子構造の直列抵抗によるものと考えられる)

☐ The difference between curves LL´ and LL can be attributed to the reemission of electrons from shallow electron traps.
(曲線LL´とLLの差は，浅い電子トラップからの再放出によるものと考えられる)

ポイント☞名詞としては「属性(本来備えている性質)・特質」．attribution にも属性の意味がある．

average

(averaged, averaged, averaging)

〈他〉（数量などを）平均する，平均して〜となる

☐ The magnetic field is averaged in the x direction.
(磁界をx方向に平均化する)

☐ We average the input admittances by taking the mean value over eight points with equal intervals of 0.125.
(0.125の等しい間隔で8個の点に対して平均値をとることで，入力アドミタンスの平均をとる)

☐ The purpose is to eliminate the random noise by averaging the stochastic process.
(この目的は，確率過程の平均をとることによってランダムノイズを除去する

ことである)

ポイント☞名詞としては「平均」，形容詞としては「平均の」の意味. average velocity(平均速度)，on (an[the]) average(平均して)

avoid

(avoided, avoided, avoiding)

〈他〉（望ましくない物・事を)避ける，回避する

avoid doing：～することを避ける

☐ Some problems associated with this correlation technique can be avoided by using a coherent detection scheme.

(この相関手法に伴う問題は,コヒーレント検波方式を用いることで回避できる)

☐ The output of the first stage of the operational amplifier is monitored by a class C amplifier in order to avoid increasing the input capacitance of the operational amplifier.

(演算増幅器の初段の出力は,演算増幅器の入力容量が増加するのを避けるために，C級増幅器によって監視される)

☐ The phase shifter is operated slightly above the Curie temperature to avoid hysteresis.

(その移相器はヒステリシスを避けるために,キュリー温度よりもわずかに高い温度で動作する)

☐ The integration in the elliptical regions is calculated by a simple formula so as to avoid time-consuming numerical methods.

(楕円領域における積分は，時間のかかる数値法を避けるために,簡単な式で計算される)

☐ The precise value of T is not important, but must be chosen large enough to avoid ringing effects.

(Tの正確な値は重要でないが,リンギングの影響を避けるだけの十分大きな

値を選ばなければならない）

☐ To avoid using matching circuits,the transistor was connected on the other side of the patch by using a probe through the substrate.
（整合回路の使用を避けるために，トランジスタは基板を貫通したプローブを用いることで，パッチの反対側で接続された）

ポイント☞［類］evade　n:avoidance 回避　adj.avoidable 避けられる
☞目的語として，(代)名詞・動名詞を伴うが，不定詞は伴わない．

β

become
(became, become, becoming)

〈自〉（物・事が）～になる
※補語は名詞・形容詞・過去分詞

☐ Statistical models of images have recently become a useful tool in image processing and computer vision.
（画像の統計モデルは最近，画像処理やコンピュータビジョンにおける有用な手段となった）

☐ The resultant feedback loops become unstable.
（結果として得られたフィードバックループは不安定となる）

☐ The problem of testing for robust stability of linear systems under uncertainty conditions has recently become of much importance and interest.
（不確実な条件のもとで線形システムのロバスト安定性を試験する問題は，最近非常に重要で，興味深いものとなった）

☐ The analysis of this electromagnetic field problem becomes simple if the Fourier-transformed domain method is used.
（この電磁場問題の解析は，フーリエ変換領域法が使用されると簡単になる）

☐ Hardware simulators are becoming increasingly common in today's VLSI design environments.
（ハードウェアシミュレータは，今日のVLSI設計環境でますます普及しつつある）

ポイント☞「～するようになる」は come to do を使い, become to do は誤り.

begin

(began, begun, beginning)

〈他〉（事を）始める
begin to do/doing：～し始める
〈自〉（事が）始まる

☐ Since this thickness corresponds to the thickness below which perpendicular coercivity begins to fall drastically, as shown in Fig. 2, the layer is considered to be the so-called initial growth layer with low coercivity.
（この厚みは，図２に示すように,これ以下では垂直保磁度が著しく低下し始めるという厚みと一致するので，層はいわゆる保磁度の低い初期成長層とみなされる）

☐ Mathcad enables the students to begin developing an appreciation for the capabilities of this type of engineering tool just as they appreciate those of their calculator.
（ちょうど計算器の機能を正しく評価するように，Mathcadによって学生はこのタイプの工学用ツールの機能に対する評価の展開を開始できる）

☐ In this paper, we begin by reviewing the basic event-driven logic simulation algorithm.
（本論文では,まず初めに基本的な事象駆動論理シミュレーションアルゴリズムを概観する）

ポイント☞［類］start, initiate　　n:beginning 始め

☞他動詞としては(代)名詞の他に，不定詞や動名詞も目的語とするが，不定詞を伴うことのほうが多い．不定詞は動作の起点に重点を置く場合，動名詞は意識的動作の継続に注意を払う場合に使われる．自動詞用法では，「～から始まる」の前置詞は from ではなく，at(時点)，on(特定の日)，in(年・月)というように前置詞を使い分ける必要がある．

behave

(behaved, behaved, behaving)

〈自〉（～に対して)ふるまう，挙動する，動作する

□ The alkyl group Ⅵ compounds behave similarly to the heterocyclic materials.
(このアルキル基Ⅵ化合物は複素環式材料と似た挙動を示す)

□ For practical purposes the noise source behaves as if an external flux noise were applied to the SQUID.
(実際の用途の場合，この雑音源はあたかも外部磁束雑音がSQUIDに加えられるかのように挙動する)

□ This filter behaves like a bandpass filter in parallel with a lowpass filter.
(このフィルタは低域フィルタと並列している帯域フィルタのような動作をする)

□ This Figure indicates that the device behaves like the Fabry-Perot etalon and can be used as a narrowband filter.
(この図は，デバイスがファブリーペロエタロンのように動作し，狭帯域フィルタとして用いることができる，ということを示している)

ポイント☞n:behavior ふるまい

☞この behavior は不可算名詞なので，形容詞がついても不定冠詞はつかない．

bring

(brought, brought, bringing)

〈他〉 (物を)持ってくる
(事を)もたらす
bring A B／B to A：AにBを持ってくる(もたらす)
bring about A／A about：Aを引き起こす，もたらす
bring down A／A down：Aを下げる，減少させる
bring in A／A in：Aを生じさせる，導入する

□ By adding a small amount of $PbTiO_3$,the Curie temperature of $Sr_{1-x}Pb_xTiO_3$ can be brought slightly below 77 to 100 K.
(少量の$PbTiO_3$を加えることで，$Sr_{1-x}Pb_xTiO_3$のキュリー温度を，やや77 Kより低い温度から100 Kへ持ってくることができる)

□ The goal of the control problem is to move the cart to a target position and bring the pendulum bob to the upright position simultaneously.
(この制御問題の目的は，カートを目標位置まで移動させ,同時に振子のおもりを垂直に持ってくることである)

□ In thc thin-film-SOI devices, three factors bring this about.
(薄膜SOIデバイスの場合，三つの要素がこれをもたらす)

□ If the capacitor consisted of a single insulating layer, the aperture ratio would be brought down to 54%.
(もしこのコンデンサが単一の絶縁層から構成されていれば,口径比は54%まで減少するのだが)

□ The nonlinear function g_p of the physical system brings in nonlinearity and causes distortion.
(その物理的システムの非線形関数g_pは非線形性を生じさせ,歪みの原因となる)

ポイント☞［反］take(持って行く)
☞bring はこちらへ持ってくる場合,take はこちらから向こうへ持っ

て行く場合に用いる．

build

(built, built, building)

〈他〉（装置などを）作る，組み立てる
（建築・構造物などを）建設する，建てる
build up：増大する，増加する

☐ The thermal controller is built in a closed-loop configuration which can be analyzed using the classical linear feedback control theory.
（熱制御装置を，古典的な線形フィードバック制御理論を用いて解析できる閉ループ構成で組み立てた）

☐ The STW delay lines were built on Y-cut crystal substrates with 25 nm thick metalization.
（STW遅延線を，25 nmの厚さのメタライゼーションを施したYカット水晶基板上に作った）

☐ The tachometer was built for a microprocessor-based system devoted to design and test dc machine controllers.
（その回転速度計は，直流機械用制御装置の設計と試験に当てられるマイクロプロセッサベース・システム用に作られた）

☐ It has been shown that this current-to-voltage converter can be built readily in any electronic lab.
（この電流－電圧変換器は，どの電子工学実験室でも簡単に作ることができることが示されている）

☐ A pilot plant was built in the laboratory to test the failure isolation presented above.
（上で示した故障分離を試験するために，パイロットプラントを実験室に作った）

☐ When the signal increases, current builds up in the junction and causes it to

send out a number of pulses.
(信号が増えると，電流が接合部で増加し，その結果,接合部は多くのパルスを出す)

ポイント☞［類］make, manufacture, fabricate, assemble, form, produce, prepare, construct　　n:building 建造物・建物
☞「計画や工程に従って製品や建造物を作る」というニュアンスがある.

C

calculate
(calculated, calculated, calculating)
〈他〉（数式などを）計算する

□ The numerical values for the trapping energy levels can be calculated from the plots of $\ln(e_n/T^2)$ vs $10^3/T$,where the trap energy levels are obtained from the slopes of these plots.
(この捕獲エネルギー準位に対する数値は, $\ln(e_n/T^2)$ vs $10^3/T$のグラフから計算できる.ここで捕獲エネルギー準位はこれらのグラフの傾きから得られる)

□ The interface temperature is calculated as a function of the drive voltage.
(界面温度は駆動電圧の関数として計算される)

□ The performance can be calculated by the following methods:
(性能は，次の方法によって計算できる)

□ We calculate the spectra of EEG data and analyse them based on the Lyapunov analysis.
(EEGデータのスペクトルを計算し,リャプノフ解析に基づいてそのスペクトルを解析する)

☐ Most engineers calculate probability of component failure using PCA or Markov analysis.
(エンジニアのほとんどは，部品の故障率をPCA，あるいはマルコフ解析を用いて計算している)

☐ The boundary element method is used to calculate the electric fields.
(境界要素法は電場を計算するのに用いられる)

☐ It is easy to calculate the coefficient matrix B provided that S is nonsingular.
(Sが非特異だとすれば，係数行列Bを計算することは容易である)

☐ These matrices are used in calculating the frequency response of the network.
(これらの行列は，回路網の周波数応答を計算する際に使用される)

☐ The permittivities for water and methanol obtained from the experiments agree well with the calculated permittivities over the entire frequency range.
(実験で得られた水とメタノールの誘電率は，全周波数領域にわたって計算された誘電率とよく一致する)

ポイント☞［類］compute　　n:calculation 計算
☞「高度で複雑な計算をする」のニュアンスがある．calculator は計算器．

calibrate
(calibrated, calibrated, calibrating)
〈他〉（計測器を）校正する

☐ The gain of the amplifier was calibrated using a lamp.
(増幅器の利得をランプを用いて校正した)

☐ To validate this procedure,the system was calibrated with three two-port devices of equivalent length.
(この方法を実証するために，長さが等しい三つの２ポートデバイスを用いてシステムを校正した)

☐ An accurate fiber optic attenuator is useful to <u>calibrate</u> optical measurement instruments such as power meters and optical time-domain reflectometers.
(精密な光ファイバ減衰器は，電力計や光時間領域反射率計のような光学測定計器を校正するのに有用である)

☐ It is seen that the standard deviation for the <u>calibrated</u> data is significantly smaller than that of the uncalibrated measurements.
(校正データに対する標準偏差は，未校正測定値のよりも著しく小さいということがわかった)

ポイント☞n:calibration 校正
☞calibration curve(校正曲線)

call
(called, called, calling)

〈他〉 (名前などを)呼ぶ，名づける
call A B：AをBと呼ぶ
〈自〉 (人が)呼ぶ
call for A：Aを必要とする，要求する

☐ The patterns in the codebook are <u>called</u> codevectors.
(コードブック中のそのパターンを，コードベクトルという)

☐ We <u>call</u> this plot the characteristic boundary curve (CBC).
(この図を特性境界曲線(CBC)と名づける)

☐ We <u>call</u> it a switching device in this paper.
(本論文では，それをスイッチング素子と呼ぶ)

☐ This situation will increasingly <u>call</u> for broadband multimedia communication.
(この状況は，広帯域マルチメディア通信をますます必要とする)

ポイント☞［類］name,designate,refer to
☞名詞としては「(電話の) 呼び出し」.

cancel

(canceled, canceled, canceling)

〈他〉（物・事を）取り消す，相殺する

☐ The nonlinear term can be <u>canceled</u> out by subtracting an estimate of it from the right-hand side of 2(a).
（非線形項は，その推定値を2(a)の右辺から引くことで消去できる）

☐ This signal <u>cancels</u> the quantization noise by adding it to the first stage output.
（この信号により，量子化雑音を第一段の出力に加えることでこの雑音は打ち消される）

☐ This perturbation will be used to <u>cancel</u> the $2\varepsilon X_n E_n \xi_n$ term in (4.6).
（この摂動は，(4.6)の$2\varepsilon X_n E_n \xi_n$項を消去するのに用いることができる）

☐ We used (5.10) to <u>cancel</u> the coefficient of A in the right side of (5.9).
（(5.9)の右辺にあるAの係数を消去するのに，(5.10)を用いた）

ポイント☞名詞としては「取り消し・相殺」．cancellation も取り消し．

carry

(carried, carried, carrying)

〈他〉（物を）携行する

carry out A：Aを実施する，実行する，行う

☐ The satellte <u>carries</u> an interferometer and measures range to the user and two angles.
(衛星は干渉計を搭載しており，ユーザーと二つの角度までの距離を測定する）

☐ Before completing fabrication, SPICE simulation was <u>carried</u> out.
（組み立てを完了する前に，SPICEシミュレーションを実施した）

☐ The modelling of the convertor in Fig.1 is carried out under the following assumptions.
(図1のコンバータのモデル化は，次の仮定のもとで行われる)

☐ Epitaxial growth was carried out by three-step low-pressure MOVPE.
(エピタキシー成長は3ステップ低圧MOVPEによって行われた)

ポイント☞carrier は運搬車(船)・搬送波．「実行する」の carry out の使用頻度は高い．

cause

(caused, caused, causing)

〈他〉 (物・事が～の)原因となる
(結果として～を)引き起こす，もたらす
cause A to do：Aに～させる，Aが～する原因となる

☐ The series resistance of the RC line is caused by the resistivity of the gate line material.
(このRC線の直列抵抗は，ゲート線材の抵抗率に起因する)

☐ Variability in the couplant causes change in sensitivity and frequency response.
(接触媒質の変異性は感度と周波数応答の変化を引き起こす)

☐ A void causes an increase in the thermal resistance between the chip and package in the region immediately near the void.
(ボイドは隣接した領域において,チップとパッケージの間の熱抵抗の増加をもたらす)

☐ The voltage difference between the source and the drain is a maximum and V_{BB} is a minimum, and it causes the large leakage current.
(ソース・ドレイン間の電圧差は最大，V_{BB}は最小なので,多量の漏れ電流を引き起こす)

☐ New technologies will cause drastic changes in the software market in the near

future.
(新技術は近い将来，ソフトウェア市場に劇的な変化をもたらすだろう)

□ The rotation of the antenna may <u>cause</u> the signal level to rise before the attenuator can be reset.
(アンテナの回転は，減衰器をリセットする前に信号レベルを上げる原因になることがある)

□ Acoustic phenomena <u>caused</u> by elastic non-linearity, such as high harmonics and shock wave generation, have been studied intensively for about 25 years.
(高調波や衝撃波の発生といった，弾性非線形性に起因する音響現象は約25年間，集中的に研究されてきた)

ポイント☞［類］bring about, give rise to
☞名詞としては「原因・理由」．「〜に･･･させる」という意味では make と同様に使役動詞だが，make とは異なり，cause は to を必要とする．

> **change**
> (changed, changed, changing)
>
> 〈他〉（物・事を）変える，変更する
> change A into B：AをBに変える
> 〈自〉（物・事が）変わる，変化する

□ The processor <u>changes</u> analog signals into digital signals.
(そのプロセッサはアナログ信号をディジタル信号に変換する)

□ Since the reflected beams are ignored, the refractive index can not <u>change</u> abruptly in the axial direction.
(反射ビームは無視されるので，屈折率は軸方向に急峻に変化できない)

□ The voltage output will <u>change</u> directly proportionally to the strength of the magnetic field.
(電圧出力は磁場の強さに正比例して変化する)

☐ It is shown that the convergence condition of the learning control in the feedback configuration does not change from the condition in an open-loop configuration.
（フィードバック構成の学習制御の収束条件は，開ループ構成の条件によって変わらないということが示される）

☐ The width of hysteresis loop can be controlled continuously by changing I_2.
（ヒステリシスループの幅は，I_2を変えることで連続的に制御できる）

ポイント☞［類］alter, convert, vary
☞名詞としては「変化・変更」．変化を表す最も一般的な語だが，「前の物とはっきり，あるいは全面的に異なった物になる」というニュアンスがある．

characterize

(characterized, characterized, characterizing)

〈他〉（物・事を）特徴づける

☐ Current industrial robots are characterized by a wide diversity of physical design configurations.
（現在の工業用ロボットの特徴は，広い多様性をもつ物理的設計構成にある）

☐ This phase shifter is characterized by low insertion loss and low SWR.
（この移相器は挿入損が小さく，SWRが低いというのが特徴である）

☐ In RF and microwave-frequency regions,networks are generally characterized by so-called scatter parameters.
（ＲＦおよびマイクロ波領域では，一般には回路網の特徴は，いわゆる散乱パラメータにある）

☐ So far MOS device behaviour has been characterized within the linear region of drain currents.
（今までのところ，MOSデバイスの挙動は，ドレイン電流の線形領域内で特徴づけられてきた）

☐ The amplifier noise performance can be characterized by measuring the amplifier output signal-to-noise ratio.
(その増幅器の雑音性能は，増幅器出力のS/N比を測定することで特徴づけることができる)

☐ The following result characterizes completely these filters.
(次の結果は，これらのフィルタを余すところなく特徴づけている)

☐ In Section Ⅳ, we discuss how to characterize chaos from a time series.
(第4節では，時系列からどのようにカオスの特徴を明らかにするかという方法について論じる)

☐ The readings from the counter are processed by a computer in order to characterize the oscillator.
(その発振器の特徴を明らかにするために，カウンタの読み取り値をコンピュータで処理する)

ポイント☞日本語の表現として，「～の特徴は…である」「～には…という特徴がある」とするほうが，自然である場合が多い．

check

(checked, checked, checking)

〈他〉（物・事を）調べる，点検する，確かめる，照合する
check A against B：AをBと照合する

☐ These results cannot be directly checked with a simulation program such as SPICE.
(これらの結果は，SPICEのようなシミュレーションプログラムでは直接確かめられない)

☐ Thus, the model is validated by comparison with moment method results which were previously cheched against experimental data.
(したがって，このモデルは，実験データと前もって照合されているモーメント法による結果と比較することで評価される)

☐ A digital oscilloscope was used to chech the ripple voltage.
(ディジタルオシロスコープをリプル電圧を調べるために用いた)

☐ These measurements are widely used in non-destructive testing to check elastic constants.
(これらの測定値は弾性定数を調べるために，非破壊試験で広く用いられる)

ポイント☞［類］investigate, examine, inspect
☞「間違いや異常がないかを，手早く調べる」のニュアンスがある．名詞としては「点検・照合」．checkpoint はチェックポイント．

choose
(chose, chosen, choosing)

〈他〉（複数個の中から～を）選ぶ，選択する
choose A as (to be) B：AをBに選ぶ
choose to do：～することに決める

☐ Lens diameter was chosen to be large enough to collimate all the light in the fiber output cone.
(レンズの直径を，ファイバの出力円錐内のすべての光を平行にするのに十分な長さに選んだ)

☐ The optical axis must be chosen to be essentially normal to the object plane.
(光軸は物体面に対して本質的に垂直になるように選ばなければならない)

☐ Diameter was chosen to match shaft inner diameter.
(シャフトの内径に合うような直径が選ばれた)

☐ Center bandwidth of the polarizer was chosen to match the optical source center bandwidth.
(光源の中心帯域幅と一致するように，偏光子の中心帯域幅は選ばれた)

☐ A Butterworth filter was chosen for the airflow meter design.
(風量計の設計のために，バターワースフィルタが選択された)

☐ However,in the present case the solution E(r) to the scalar Helmholtz equation can be chosen as a real function.
(しかしながら，この場合,スカラーヘルムホルツ方程式の解E(r)を実関数に選ぶことができる)

☐ Nevertheless, most researchers have adopted this approximate model, and we choose to use this model as well in this paper.
(それにもかかわらず,たいていの研究者はこの近似モデルを採用しているので，本論文でもまた，このモデルを用いることに決める)

☐ In the rest of the proofs of Lemmas 1 and 2, we choose to leave the terms of initial conditions out of the functions for simplicity.
(補助定理１と２の証明の残りにおいて,簡略化のために関数の中から初期条件の項を除くことに決める)

ポイント☞［類］select, prefer, pick　　n:choice 選択
☞「二つ以上の物から自分の判断で一つを選ぶ」のニュアンスがある.

clarify
(clarified, clarified, clarifying)
〈他〉（考え方・理論などを)明らかにする，明確にする

☐ Some general properties of the modeled CMOS processors will now be clarified.
(モデル化されたCMOSプロセッサの一般的特性が今明らかになる)

☐ Using a two-dimensional device simulation, we clarified why HEMT's have superior low-noise performancc.
(二次元デバイスシミュレーションを用いることで,HEMTがなぜ優れた低雑音性能をもつかを明らかにした)

☐ We describe a simple example, which will be used in the following sections to clariy the introduced general framwork.
(簡単な例について述べるが,この例は導入された一般的枠組を明確にするた

めに，以下の節で用いられる)

☐ We solve the problem by clarifying the ambiguity in the interactive design process.
(会話形設計過程に存在するあいまいさを明確にすることでこの問題を解く)

ポイント☞［類］disclose, reveal, elucidate　　n:clarificaton 解明
☞「はっきりしないあいまいな点をはっきりさせる」のニュアンスがある.

classify

(classified, classified, classifying)

〈他〉（物·事を）分類する
classify A into [in,as] B：AをBに分類する

☐ These features can be classified into three classes.
(これらの特徴は三つのクラスに分類できる)

☐ Lenear synchronous motors with permanent magnets may be classified into the short-permanent-magnet(PM) poles type and long PM poles type according to their structural features.
(永久磁石を有するリニア同期電動機は，構造上の特徴によって短永久磁石(PM) 極型と長PM極型に分類できる)

☐ Sensors employing an electrochemical mode of detection may be classified as either potentiometric or amperometric/voltametric.
(検出のうちの電気化学的モードを利用するセンサは，電位差測定型，または電流測定型／電圧測定型に分類できる)

☐ From a modelling point of view, it is possible to classify predictive control schemes into two main categories.
(モデリングの観点から見れば，予測制御方式は二つの主なカテゴリーに分類することができる)

ポイント☞［類］group, divide, fall into, sort　　n:classification 分類
☞「ものをカテゴリーに従って体系的に分類する」のニュアンスがある．
☞decimal classification（十進分類法）

coat
(coated, coated, coating)
〈他〉（塗料などを）塗る，塗布する，被覆する
coat A with B：AにBを塗る

☐ The surface is coated with a thin film (about 7-10 nm thick) of gold-palladium before the SEM examination.
（SEM検査の前に，表面は金－パラジウム薄膜（およそ7～10 nm厚）で被覆される）

☐ In order to facilitate this bond the fibers were coated with a thin layer of Cr/Pt/Au.
（この接合を容易にするために，ファイバはCr/Pt/Auの薄い層を塗布された）

☐ An antireflection layer was coated onto the back surface.
（反射防止層が裏面に塗布された）

☐ Figure 6 shows the pressure sensitivity of a fiber coated with Ni.
（図 6 はNiが塗布されたファイバの感圧性を示している）

ポイント☞名詞としては「塗(皮)膜・層」．coating はコーティング・被覆．

coincide
(coincided, coincided, coinciding)
〈他〉（性質などが）一致する
coincide with A：Aと一致する

☐ The phase measurements strongly coincide with the theoretical expectations.
（その位相の測定値は，理論的期待値と非常によく一致する）

☐ In this case, ALADIN cioncides with the EBP algorithm.
（この場合，ALADINはEBPアルゴリズムと一致する）

☐ Moreover, the results nearly coincide with the values derived from the magnetic wall interface in case of $h_2 = h_1$.
（そのうえ，この結果は，$h_2 = h_1$の場合に磁壁の境界面から導き出した値とほぼ一致する）

ポイント☞［類］agree, accord, conform, concur, correspond　n:coincident 一致　adj:coincident, coincidental 一致した　adv:coincidentally 一致して
☞「性質・意見などが完全に一致する」のニュアンスがある．

collect
(collected, collected, collecting)

〈他〉（信号・データなどを）集める，収集する
〈自〉（信号・データなどが）集まる，たまる

☐ In reality, the input-output data sequences are collected by measurement.
（実際は，入出力データ列は測定することで収集される）

☐ Data were collected using a 16 bit A/D converter.
（データは16ビットA/D変換器を用いて収集された）

☐ The light from the source is collected by collection lens CL.
（光源から出た光は，集光レンズCLによって集められる）

☐ The backward travelling light is collected, at the fiber input end, through a beam splitter.
（後進進行光はビームスプリッタを介して，ファイバの入力端で集められる）

☐ When light is incident on the detector it induces a photocurrent which is collected by the electrodes.
（光が検出器に入射すると，電極によって集められる光電流を誘導する）

☐ The signals collected at the subsequent transducer positions are stored after an A/D conversion in a buffer memory and analysed by a computer.
（トランスデューサの次の位置で集められた信号は，A／D変換の後でバッファメモリに格納され，コンピュータで解析される）

ポイント☞［類］gather　n:collection 収集　adj:collective 集めた
☞「はっきりとした目的のために注意深く集める」のニュアンスがある．

combine

(combined, combined, combining)

〈他〉（物・事を）結合させる，組み合わせる
combine A with B：AをBと組み合わせる
combine A and B：AとBを兼ね備える

☐ We combined this servo with the 2-D tracker circuit on a single chip.
（シングルチップ上にこのサーボを2-Dトラッカ回路と組み合わせた）

☐ The technique combines optical lithography and spatial filtering.
（その手法はフォトリソグラフィと空間フィルタリングとを兼ね備えている）

☐ Multiple-layer neural network can be made by combining two or more neuro chips.
（2個以上のニューロチップを組み合わせることによって，多層ニューラルネットワークが作れる）

ポイント☞［類］join, bond, couple, unite　n:combination
☞「それぞれの特性を保ったまま結合させる」のニュアンスがある．

come

(came, come, coming)

〈自〉（物が）来る，着く，達する，現れる
（事が）起こる
come from A：Aより生じる，Aに起因する

come to do：～するようになる

☐ This noise comes from the amplification which takes place before the conversion.
（この雑音は変換前に行われる増幅より生じる）

☐ Electromagnetic noise comes primarily from the large electric motors used to turn the mill.
（電磁雑音は主として，ミルを回転させるのに用いられる大型電動機に起因する）

☐ Rules have to be extracted and formalized in order to come to appropriate knowledge-based CAD tools suited to support these tasks.
（これらのタスクを支援するのに適した知識ベースCADツールを使用できるようになるには，ルールを抽出し，公式化しなければならない）

☐ The light coming from the light source is reflected towards the screen.
（光源から来る光はスクリーンに向かって反射する）

☐ Work is in progress to solve the problems coming from the bandwidth reduction due to the parasitic capacitance.
（寄生容量による帯域幅縮小に由来する問題を解決するための研究が進行中である）

ポイント☞ “名詞＋to come” の形の場合は「将来（未来）の～」の意味. for a year to come(今後 1 年間)

communicate

(communicated, communicated, communicating)

〈他〉（情報・データなどを）伝達する
〈自〉（人・物が）通信する
communicate with A：Aと通信する

☐ For the linear network of processors that are equipped with front end processors,the processors can communicate and compute at the same time.

（フロントエンドプロセッサが実装されているプロセッサからなる線形回路網の場合，プロセッサは通信と計算が同時にできる）

☐ The card communicates with the PC via the digital interface.
（そのカードはディジタルインタフェースを介してPCと通信する）

☐ The RS3 system monitored the process and communicated with a PLC.
（ＲＳ３システムはプロセスを監視し，PLCと通信した）

☐ The digital section is concerned with processing and communicating the data.
（ディジタルセクションはデータを処理し，伝達することにかかわっている）

ポイント☞n:communication 通信
☞人を目的語にとらない．digital communication(ディジタル通信), data communication(データ通信), optical communication(光通信)

compare

(compared, compared, comparing)

〈他〉（物・事と）比較する

compare Ａ with（to）Ｂ：ＡをＢと比較する

☐ The thicknesses obtained by the these methods are compared in Fig.3.
（これらの方法で得られた厚さを図３で比較する）

☐ The experimental results are compared with simulation results in Fig.1.
（実験結果を図１のシミュレーション結果と比較する）

☐ Numerical results are presented and compared with exact solutions for homogeneous and inhomogeneous circular cylinders.
（数値結果を示し，均質および不均質な円柱に対する厳密解と比較する）

☐ The filter output is compared to a DC reference voltage by the servo amplifier.
（フィルタの出力をサーボ増幅器による直流基準電圧と比較する）

☐ We compare these parameters with the value expected from the physical and material data.
(これらのパラメータを物理的データと物質的データからの期待値と比較する)

☐ To assess the merits and restrictions of each method we compare the spectral-domain model to the lumped-parameter model.
(それぞれの方法の利点と制約を評価するために，スペクトル領域モデルを集中定数モデルと比較する)

☐ Compared Fig.3 with Fig.2, peak acoustic pressure decreases to 1/3 and rise time increases to 400 ns.
(図３を図２と比較すると，最大音圧は1/3に減少し，立ち上がり時間は400 nsまで増加している)

☐ Such patterns can be detected by comparing the image with templates.
(そのようなパターンは，画像をテンプレートと比較することで検出できる)

ポイント☞n:comparisn 比較　adj:comparable 比較できる，comparative 比較による　adv:comparatively 比較的
☞to は受動態で好まれる．

compensate
(compensated, compensated, compensating)
〈他〉（損失などを）補償する，補う

☐ Even though the plant is a mechanical system,the strong instability of the open-loop plant can be compensated with a phase-lead characteristics of the controller.
(たとえプラントが機械系であっても，開ループプラントの強い不安定性は制御装置の位相進み特性によって補償できる)

☐ The base current of the transistor T_1 is compensated by the transistor T_3. Transistor T_5 compensates the base current of the switching transistor S_1.

(トランジスタT_1のベース電流はトランジスタT_3によって補償される．トランジスタT_5はスイッチングトランジスタS_1のベース電流を補償する)

□ Employing conventional feedback controllers can be helpful to overcome this kind of problem in the transient stages of learning since they can compensate the control input to reduce the error.
(従来のフィードバック制御器の使用は，制御入力を補償し,誤差を低減できるので，学習の過渡段階で生じるこの種の問題を解決する助けになる)

□ This will require the use of sophisticated multimedia end systems which are able to compensate for the deficiencies of the networks.
(これには,ネットワークの欠陥を補うことができる精巧なマルチメディア端末装置を使用することが必要である)

□ Phase shifters are used to compensate the auxiliary phase shifts caused by the electronic components used.
(用いられた電子部品が原因で起こる補助機器の位相ずれを補償するために，移相器を使用する)

ポイント☞［類］make up for　　n:compensation 補償
☞compensator は補償器．feedback compensation(フィードバック補償)

compose
(composed, composed, composing)
〈他〉（物を）組み立てる，構成する
be composed of Ａ：Ａからなる，Ａで作られている

□ This circuit is simple,and it is composed of a diode, a capacitor and a resistor.
(この回路は簡単で，ダイオード１個，コンデンサ１個,抵抗器１個で作られている)

□ In this report, we shall examine that Electroencephalogram (EEG) may be composed of some frequency components with chaotic characteristics.
(この報告書では，脳電図(EEG)はカオス的特性のある周波数成分からなる)

☐ The matrix model that represents the converter system is composed of three elements:
(この変換器システムを表現する行列モデルは３個の要素からなる)

ポイント☞［類］constitute　　n:composition 構成
☞be composed of というように，受動態で使われることが多い．

compress
(compressed, compressed, compressing)
〈他〉（信号・データ・気体などを）圧縮する

☐ Radar pulses must be compressed by signal processing in the receiver.
(レーダパルスは受信機での信号処理によって圧縮されなければならない)

☐ The output of a combinational logic circuit with *n* inputs and *m* outputs, $n>m$, is compressed with the ratio of m/n.
(入力がn個，出力がm個で，$n>m$の組み合わせ論理回路の出力は，m/nの比で圧縮される)

☐ Because air is being compressed for combustion within the engine, the recirculation of this air can reduce horsepower and efficiency.
(空気はエンジン内の燃焼のために圧縮されているので，この空気の再循環で馬力と効率は低下することもある)

☐ Vector quantization can effectively compress images using categorization.
(ベクトル量子化は，カテゴリー化を用いて画像を効果的に圧縮できる)

☐ As a result, a controlling signal is generated that is used to set the gain of the amplifier necessary in order to compress the dynamic range.
(その結果，ダイナミックレンジを圧縮するのに必要な増幅器の利得を設定するのに用いられる制御信号が発生する)

☐ This paper derives an optimum quantization scheme for compressing the aperture data in a microwave antenna array.

（本論文では，マイクロ波アンテナアレイにおける開口データを圧縮するための最適量子化方式を導き出す）

ポイント ☞n:compression 圧縮

☞compressor は圧縮機・コンプレッサー．compressed air(圧搾空気・圧縮空気)

compute

(computed, computed, computing)

〈他〉（数・量などを）計算する

□ The magnetic field distribution for a thin magnetic film is computed using the fast Fourier transform technique.

（磁性薄膜に対する磁場分布を高速フーリエ変換法を用いて計算する）

□ The algorithm presented here computes the autoregressive spectrum directly from data.

（ここで示したアルゴリズムは，自己回帰スペクトルをデータから直接計算する）

□ In the most general case, one computes the spectral energy distribution and then converts it to the signals necessary to drive the display device.

（最も一般的な場合，スペクトルエネルギー分布を計算し，次に表示装置を駆動するのに必要な信号にその分布を変換する）

□ These currents can be used to compute the scattered electric and magnetic fields in the region exterior to the scatter.

（散乱体から離れている領域の散乱電場と磁場を計算するのに，これらの電流を用いることができる）

□ Examples of recently computed results based on the complete Navier-Stokes equations are shown in Fig.1.

（完全ナヴィエーストークス方程式に基づいて最近計算された結果の例を図1に示している）

□ The flow graph for computing these transform are shown in Fig.2.
（これらの変換を計算するためのフローグラフが図２に示されている）

ポイント☞［類］calculate　　n:computation 計算
☞「簡単な，あるいは規模の大きい計算をする」のニュアンスがある．computer はコンピュータ．

concern
(concerned, concerned, concerning)
〈他〉（物・事が）～に関係がある，～に関する，～にかかわる
be concerned with Ａ：Ａと関係する
be concerned about Ａ：Ａを心配している

□ This paper is concerned with position errors along the z-axis.
（本論文は，z軸に沿った位置誤差に関するものである）

□ We are mainly concerned about the high frequency vertical vibrations.
（高周波垂直振動を主に心配している）

□ Section Ⅲ concerns the question of how to determine an optimal approximation of an observed time series within a given model.
（第２節は，所定のモデル内で観測された時系列の最適近似をどのようにして決定するかという問題に関係している）

□ EMC problems concerning thermometric systems for electromagnetic hyperthermia are discussed.
（電磁温熱療法用の温度測定システムにかかわるＥＭＣ問題について論じる）

ポイント☞［類］be related to,be associated with,refer to
☞名詞としては「関係」．as (so) far as～be concerned は「～に関する限り」で，通例文頭で用いる．

conduct

(conducted, conducted, conducting)

〈他〉（研究・実験などを）行う，実施する

□ Frequency domain analysis was <u>conducted</u> using an HP8505A network analyzer.
（周波数領域解析をHP8505Aネットワークアナライザを用いて行った）

□ All the above experiments were <u>conducted</u> assuming $v(n)=0$.
（上記の実験はすべて$v(n)=0$と仮定して実施した）

□ Extensive measurements were <u>conducted</u> on the antenna (Fig.1) to find the parameters which minimize the operating frequency.
（動作周波数を最小にするパラメータを見いだすために，アンテナ（図１）に対して広範な測定を実施した）

□ Using this plasma source, we <u>conducted</u> several etching experiments for the basic evaluation method for processing.
（このプラズマ源を用いて，プロセシングの基礎的評価法のために，いくつかのエッチング実験を行った）

ポイント☞［類］do, perform, make, run
☞名詞としては「行い・実施」.

confirm

(confirmed, confirmed, confirming)

〈他〉（証拠・予測などを）確かめる，立証する，確認する
confirm that節：～ということを裏づける，確認する

□ The high quality of the films was <u>confirmed</u> by every device having a resistance in excess of 30 MΩ.
（その膜が高品質であるということを，30 MΩを越える抵抗を有するあらゆる装置によって確認した）

□ We <u>confirmed</u> the synchronization phenomena by circuit experiments and

numerical calculations.
(その同期現象を回路実験および数値計算によって確認した)

☐ Simulation results which confirm the analysis are also given.
(その解析を立証するシミュレーション結果も与えた)

☐ The results of X-ray analysis confirm that coating is of copper only.
(X線分析の結果により，コーティングは銅のみからなるということを確かめた)

ポイント☞［類］ascertain,verify,make sure　　n:confirmation 確認
☞「論じられたことが正しいということを確かめる」のニュアンスがある．

connect
(connected, connected, connecting)

〈他〉（二つ以上の物を）つなぐ，接続する，連結する
connect A to (with) B：AをBと接続する
〈自〉（二つ以上の物が）つながる，接続する，連結する

☐ In order to determine the stability of the data acqusition unit a 2.5-MHz crystal controlled oscillator was connected directly to the mixer stage.
(データ収集装置の安定性を決定するために，2.5MHz水晶制御発振器をミキサー段に直接接続した)

☐ The upper electrode is connected to a 100 kHz rf oscillator through a 47 nF capacitor.
(上部電極は47 nFコンデンサを介して100 kHz高周波発振器と接続される)

☐ The two coils are connected in parallel with each other and with the input coil of a SQUID.
(二つのコイルは並列接続され，さらにSQUIDの入力コイルに接続される)

☐ If we connect the points that are equally distant from Q_0 and L, we can divide

the space into two regions as shown in Fig.1.
(もしQ_0とLから等しい距離にある点を結ぶと，図１に示すように，空間を二つの領域に分けることができる)

□ These actuators may <u>connect</u> directly to the mechanical links.
(これらのアクチュエータは機械的リンクと直接連結できる)

□ The resistors and the operational amplifier <u>connected</u> to the taps of the delay circuit form a summing circuit.
(遅延回路のタップと接続されている抵抗器と演算増幅器は，和回路を構成する)

ポイント☞［反］disconnect(切断する)　　n:connection 接続・連結
☞connector はコネクタ．

consider
(considered, considered, considering)

〈他〉（問題などを）よく考える，考察する
consider A（to be）B：AをBだとみなす
consider that節：～だと考える

□ Three different types of refractive index distributions are <u>considered</u> in this section.
(この節では，３種類の屈折率分布について考察する)

□ The element values are <u>considered</u> as constant except for the transconductance g_m.
(素子の値は相互コンダクタンスg_m以外は定数と見なされる)

□ We <u>consider</u> in this paper an induction heating process.
(本論文では誘導加熱の過程について考察する)

□ Many designers of Home Automation Systems seem to <u>consider</u> a PC the most convenient control device for their networks.

(ホームオートメーションシステムの設計者の多くは，PCはネットワークの最も都合のよい制御装置であるとみなしているらしい)

☐ When $N=9$, we considered that we could not see 9-phase oscillation.
($N=9$の場合，9相発振が見られなかったと考えた)

☐ In this paper, we describe a new three dimensional analysis technique that is used to obtain more accurate propagation constants by considering the E_z, H_z components of the electromagnetic field.
(本論文では，電磁場のE_z，H_z成分を考慮することで，より正確な伝搬定数を得るために用いられる，新しい三次元解析手法について述べる)

☐ Considering the above limitations,a parallel systolic architecture is not an appropriate choice for this purpose.
(上記の制限を考慮すれば,並列シストリックアーキテクチャはこの目的には適切な選択ではない)

ポイント☞ ［類］think　n:consideration 考察　adj:considerable かなりの・重要な

☞consider～ing は「～することを考える」, consider wh-節・句は「～かどうかを考える」.

consist

(consisted, consisted, consisting)

〈自〉（部分・要素から）なる

consist of A：Aからなる

☐ As shown in Figure 1,the Wheatstone bridge consists of six resistors of equal value.
(図1に示すように,このホイートストンブリッジは値が等しい6個の抵抗器からなる)

☐ The problem of failure detection consists of detecting failures in a physical system by monitoring its inputs and outputs.

（この故障検出問題は，その入力と出力を監視することで,物理システムの故障を検出することからなる）

☐ Since the antennas are small and consist of a single turn,the sensor must be operated at a frequency high enough to ensure an adequate signal level.
（このアンテナは小さく，そしてひと巻からなっているので,センサは十分な信号レベルを確保するのに足りる高い周波数で動作しなければならない）

☐ The method consisted of transmitting a short pulse of light through the fiber and measuring the time dependence of the Rayleith backscattered light.
（この方法は，ファイバを通じて短い光パルスを伝送し,レーリー後方散乱光の時間依存性を測定することからなる）

☐ To improve the linearity a second statge, consisting of T_3, T_4, R_3, and R_4, has been added.
（線形性を向上させるために，T_3, T_4 ,R_3, R_4からなる二段が追加された）

ポイント☞［類］be composed of, be made up of, comprise
☞consist は自動詞なので，be consisted of は誤り．また,進行形は不可．

construct

(constructed, constructed, constructing)

〈他〉（部品などを)組み立てる

☐ The yokes were constructed from 0.65 mm thick steel laminations.
（その継鉄は0.65 mm厚の鋼製ラミネーションから組み立てられた）

☐ An automatic blood pressure controller is constructed using fuzzy set theory.
（自動血圧制御装置をファジィ集合理論を用いて組み立てる）

☐ The discharge chamber was constructed by machining a stainless steel tube and was mounted on a table.
（放電チャンバはステンレス鋼管を機械加工することで組み立てられ,テーブ

ルの上に取り付けられた)

□ In this section we first construct the filtering model by the reference probability method.
(この節では，初めに標準確率法によってフィルタリングモデルを構築した)

ポイント☞［類］build, make, manufacture, fabricate, assemble, form, produce, prepare n:construction 建設・構造
☞「設計に従って部品などを組み立てる」のニュアンスがある.

consume
(consumed, consumed, consuming)
〈他〉（電力などを）消費する

□ Traditional overload relays consume 6-8 watts of electricity;the new devices use less than 0.2 watts.
(従来の過負荷継電器は6～8ワットの電力を消費する．この新しい装置は0.2ワット以下しか消費しない)

□ Under normal operating conditions,the synchronous oscillator consumes less than 1 mW.
(正常動作時では，この同期発振器の消費電力は1 mW以下である)

□ At the same time, the shift register should consume little power.
(同時に，このシフトレジスタはほとんど電力を消費しないはずである)

□ The chip operates from a power-supply voltage of 2.7 to 3.6 V, and consumes less than 1 W of power when decoding audio and video signals.
(そのチップは2.7～3.6Vの電源電圧で動作し，オーディオ信号とビデオ信号を復調するときに，1W以下しか消費しない)

ポイント☞n:consumer 消費者，consumption 消費
☞power comsumption(電力消費)

contain

(contained, contained, containing)

〈他〉（物を）含む

☐ The computation procedure for designing a robust control is contained in Section Ⅲ.
（ロバスト制御を設計するための計算手順は第3節に含まれている）

☐ It is well recognized that digital video signals contain a large number of data.
（ディジタルビデオ信号には大量のデータが含まれているということは，十分認識されている）

☐ It is now common practice to design photodiodes that contain hetero interfaces within the depletion region.
（空乏領域内にヘテロ界面を含んでいるフォトダイオードを設計することが，今や普通になっている）

☐ Most of these systems contain various nonlinearities.
（これらのシステムのほとんどには，さまざまな非線形性が含まれている）

☐ The chip area is 2240×2240 μm^2,and it contains 12 MESFETs, 14 resistors, and 16 capacitors.
（そのチップの面積は2240×2240 μm^2で，12個のMESFET，14個の抵抗器，16個のコンデンサが含まれている）

☐ A tunnel diode is a small two-terminal device containing a single junction formed by heavily doped semiconductor materials.
（トンネルダイオードとは，多量にドープされた半導体材料で形成された一つの接合を含んでいる，小さな二端子素子のことである）

ポイント☞［類］include, involve　　n:content 内容物・含有量
☞「ある物の中に内容物・成分を含む」のニュアンスがある．進行形は不可．container は容器・コンテナ．

control

(controlled, controlled, controlling)

〈他〉（物・事を）制御する

☐ The system is controlled by a 16bit, 20MHz host computer.
（そのシステムは16bit，20MHzホストコンピュータで制御される）

☐ The gain of the circuit is controlled by the bias current of each differential amplifier.
（回路の利得は，各々の差動増幅器のバイアス電流によって制御される）

☐ We assume that the plant to be controlled is a SISO,continuous time system.
（制御されるプラントは，SISOの連続時間システムだと仮定する）

☐ The attenuator connected to the multiplier can controls the amplitude of the modulated signal.
（乗算器に接続されたこの減衰器は，被変調信号の振幅を制御できる）

☐ The frequency and amplitude of the power line signal were measured,and these measurements were used to control the synthesizing sine wave generator.
（電力線信号の周波数と振幅を測定し，これらの測定値を合成正弦波発生器を制御するのに用いた）

☐ As shown in Fig.7, the controller is essentially a sequential finite state machine controlled by a two-phase clock.
（図７に示すように，この制御装置は本質的に，２相クロックで制御される順序有限状態機械である）

☐ It is commonly used for controlling the speed, position, or torque of dc motors and stepper motors.
（それは，直流電動機とステッパー電動機の速度や位置，またはトルクを制御するために一般に用いられる）

☐ Controlling the bias current is the most effective means to reduce the power

consumption.
(バイアス電流を制御することは,電力消費を少なくするための最も効果的な方法である)

ポイント☞名詞としては「制御」だが,可算名詞の用法では「制御装置」. controller は制御装置・制御器・コントローラ. automatic control (自動制御), feedback control(フィードバック制御), process control (プロセス制御)

converge
(converged, converged, converging)
〈自〉(級数・アルゴリズムなどが)収束する

☐ The FIR canceller quickly converges to its steady-state value 22 dB.
(ＦＩＲキャンセラは定常値である22 dBに急速に収束する)

☐ Fourier coefficients of the capacitance between receiving and transmitting electrodes should converge to zero as rapidly as possible.
(受信用電極・送信用電極間の静電容量のフーリエ係数は,できるだけ速く零に収束しなければならない)

☐ It has been shown that this algorithm converges to the eigenvector associated with the smallest eignvalue of the correlation matrix.
(このアルゴリズムは,相関行列の最小固有値に関連する固有ベクトルに収束するということが示されている)

ポイント☞［反］diverge(発散する)　n:convergence 収束　adj:convergent 収束の

convert
(converted, converted, converting)
〈他〉(信号などを)変換する
convert A into (to) B：AをBに変換する

☐ These digital signals are converted to analog signals by D/A converters.

(これらのディジタル信号はD/A変換器によってアナログ信号に変換される)

☐ The duty cycle is converted to a dc voltage by low-pass filtering.
(デューティサイクルは低域フィルタリングによって直流電圧に変換される)

☐ This output voltage is converted into a digital code for further data processing.
(さらにデータ処理をするために,この出力電圧はディジタルコードに変換される)

☐ The output voltage of the class C amplifier is converted to current by the V-I converter.
(C級増幅器の出力電圧は，V-I変換器によって電流に変換される)

☐ A low-pass filter converts the duty cycle information into a voltage.
(低域フィルタはデューティサイクルの情報を電圧に変換する)

☐ This circuit converts the optical signals to electrical signals.
(この回路は光信号を電気信号に変換する)

☐ To convert the nonlinear capacitance change into linear digital output, two techniques have been proposed with satisfactory results.
(非線形容量の変化を線形ディジタル出力に変換するために,満足のいく結果をもたらす二つの手法が提案されている)

☐ This converted current can be controlled by inserting the resistor between MN1 and MP3.
(この変換された電流は，MN1とMP3の間に抵抗器を挿入することで制御できる)

☐ The phase between two waves of the same frequency is computed by first converting the waves into two square waves and then measuring the time difference between the pulse centers of these square waves.
(周波数が同じ二つの波動間の位相は，初めに波動を二つの方形波に変換し，それから方形波のパルス中央間の時間差を測定することで計算される)

□ Their corresponding S-parameters can be obtained by <u>converting</u> [A^T] to [S^T].
（それらに対応するSパラメータは，[A^T]を[S^T]に変換することで得られる）

ポイント☞［類］transform, translate, transduce　　n:conversion 変換
☞「用途・目的に合うように変換する」のニュアンスがある．
converter は変換器・コンバータ．

cool

(cooled, cooled, cooling)

〈他〉（物を）冷却する，冷やす

□ The samples were first <u>cooled</u> down to 4.2 K.
（最初に，試料は4.2 Kに冷却された）

□ One can <u>cool</u> the antenna to this temperature with the aid of a large dilution refrigerator.
（大型希釈冷凍機を使って，アンテナをこの温度に冷すことができる）

□ Liquid helium is needed to <u>cool</u> these superconductors.
（これらの超伝導体を冷却するのに，液体ヘリウムが必要である）

ポイント☞［反］heat, warm(暖める)
☞形容詞としては「冷たい」．cooler はクーラー・冷却器, coolant は冷却剤．water cooling(水冷)，air cooling(空冷)．

correct

(corrected, corrected, correcting)

〈他〉（誤りなどを）訂正する，修正する
（装置・回路などを）補正する
correct A for B：BについてAを修正する

□ The back reflections were <u>corrected</u> for diffraction effects in the frequency domain.
（周波数領域における回折効果について，背面反射を修正した）

☐ This can be corrected using appropriate compensation circuits.
（これは適切な補償回路を用いることで補正できる）

☐ The attenuation coefficient estimated with the CF method was used to correct the backscattered spectrum.
（CF法によって推定された減衰係数は，後方散乱スペクトルを修正するために用いられた）

☐ This value is used to correct the phase measured with the interferometer.
（この値は干渉計で測定された位相を補正するのに用いられる）

ポイント ☞n:correction 訂正　　adv:correctly 正確に
☞形容詞としては「正しい・正確な」. error-correcting code(誤り訂正符号)

correspond
(corresponded, corresponded, corresponding)

〈自〉（物・事が）一致する，相当する
correspond with（to）A：Aに一致する
correspond to A：Aに相当する

☐ The state *x* of the model corresponds with the degrees-of-freedom.
（モデルの状態 *x* は自由度に一致する）

☐ In the shift resister case, the state of the trellis encoder corresponds to the content of the shift register.
（シフトレジスタの場合，トレリス符号器の状態はシフトレジスタの内容と一致する）

☐ We have verified dry etching characteristics with uniform etching speed distribution corresponding to the plasma distribution.
（プラズマ分布に相当する均一なエッチング速度分布を有するドライエッチング特性を確認している）

ポイント☞［類］agree, accord, conform, coincide, concur　n:correspondence 一致・相当　adj:correspondent 一致する・対応する

☞「細部にわたってではなく，全般的に一致する」のニュアンスがある．

couple

(coupled, coupled, coupling)

〈他〉（物を）結合する，連結する

couple A to B：AをBにつなぐ，結合させる

□ The transmitted light was <u>coupled</u> to the spectrometer through another fiber.
（透過光は，別のファイバを通して分光計に結合された）

□ As the manipulator system is a highly <u>coupled</u> nonlinear system,this linear model is not sufficient to describe the nonlinear couplings.
（マニピュレータシステムは高度に結合された非線形システムなので，この線形モデルは非線形結合を記述するには不十分である）

□ The presence of metal will <u>couple</u> a signal into the receivers.
（金属があると，信号は受信機につながれる）

□ Thus, we are able to <u>couple</u> control systems to the helicopter model.
（したがって，制御システムをヘリコプタモデルと結合できる）

ポイント☞［類］join, bond, combine, unite

☞「同じ類の物が二つ結合する」のニュアンスがある．coupler はカプラ・連結器．

cover

(covered, covered, covering)

〈他〉（物を）覆う，（問題などを）扱う

（範囲を）カバーする

□ The devices were <u>covered</u> with a 1-μ m CVD SiO_2 layer.
（デバイスは，1-μ m CVD SiO_2層で被覆された）

□ The temperature was varied from 77 to 300 K and the frequency covered the range of 100 Hz-several kHz.
(温度は77 Kから300 Kであり，周波数は100 Hz～数 kHzの範囲をカバーした)

□ This model covers a wide range of problems.
(このモデルは広範囲の問題をカバーする)

□ Electromagnetic interference covers the frequency spectrum from dc to the optical frequencies (30,000 GHz).
(電磁妨害は直流から光学的振動数(30,000 GHz)までの振動数スペクトルを覆っている)

□ A dielectric-coated reflector antenna consists of a metal reflector covered with a layer of dielectric material.
(誘電体被覆反射形アンテナは，誘電材料の層で覆われた金属反射器から構成されている)

ポイント☞［反］uncover(覆いを取る)
☞名詞としては「覆い・カバー」.

D

deal
(dealt, dealt, dealing)
〈自〉（～を)扱う，処理する，論じる

□ H_p fluctuations are dealt with in the frame of the theory of stochastic processes.
(H_p変動は確率過程の理論の枠組において扱われる)

☐ This paper <u>deals</u> with the design of two speech coding systems.
（本論文では，二種類の音声符号化システムの設計について扱う）

☐ This article <u>deals</u> with the relationship between Cl_2 gas corrosion and outgassing of various surface-treated aluminum alloys.
（この記事では，さまざまな表面処理を施したアルミニウム合金の塩素ガスによる腐食とガス放出の間の関係について扱う）

☐ Since we are <u>dealing</u> with digital circuits,the parasitic couplings among the partitioned subcircuits are neglected.
（ディジタル回路を取り扱っているので，分割された部分回路間の寄生結合は無視する）

☐ The first topic <u>dealt</u> with in this paper is the MOS transistor small-signal behavior, which is useful in determing circuit frequency response.
（本論文で論じられる最初の論題はMOSトランジスタの小信号挙動であり，この挙動は回路の周波数応答を決定する際に役立つ）

ポイント☞［類］treat, manage, handle
☞事柄を扱う場合は with をとるが，商品を扱う場合は in をとる．

decrease
(decreased, decreased, decreasing)
〈他〉（数・量などを）減少させる，減らす
〈自〉（数・量などが）減少する，減る

☐ It should be noticed that λ <u>decreases</u> as substrate temperature is <u>decreased</u> and becomes zero at about 200℃.
（基板温度が低下するにつれて λ は減少し，約200℃で零になるということに注意を払わねばならない）

☐ A quench current <u>decreases</u> rapidly with the increase of frequency.
（クエンチ電流は周波数の増加にともなって急速に減少する）

☐ The critical current density of the thin films <u>decreases</u> with increasing magnetic field.
（薄膜の臨界電流密度は磁場の増加とともに減少する）

ポイント☞［反］increase(増加する)　　［類］diminish, lessen, reduce
☞「数・量が次第に減少する」のニュアンスがある．名詞としては「減少」．on the decrease(次第に減少して)

define
(defined, defined, difining)
〈他〉（語などを）定義する
define A as B：AをBと定義する

☐ Stability itself has been <u>defined</u> in many different ways.
（安定性それ自体は，いろいろ異なるやり方で定義されてきた）

☐ Multimedia is <u>defined</u> as the combination of computer data, sound, animation and video.
（マルチメディアはコンピュータデータ，音声，アニメーション，ビデオを組み合わせたものと定義される）

☐ The gain of an antenna can be <u>defined</u> as a function of the scan angle.
（アンテナの利得は走査角の関数として定義できる）

☐ He <u>defined</u> the electrokinetic momentum vector A as the momentum per unit charge.
（彼は界面動運動量ベクトルAを単位電荷当たりの運動量と定義した）

☐ To do this, we <u>define</u> the acoustic impedance Z_a as the ratio of the finite Fourier transforms of sound pressure $p(t)$ and acoustic volume velocity $w(t)$.
（これを行うために，音響インピーダンス Z_a を音圧$p(t)$および音響体積速度$w(t)$の有限フーリエ変換の比と定義する）

ポイント☞n:definition 定義　　adj:definite 一定の・明確な

degrade

(degraded, degraded, degrading)

〈他〉（数・量などを）減少させる
（性能・品質などを）低下させる，劣化させる

□ As a current of amplifier is decreased, high frequency characteristic of a transistor is degraded and gain of the amplifier is decreased.
（増幅器の電流が増加するにつれてトランジスタの高周波特性は劣化し，増幅器の利得は低減する）

□ It has been reported that the maximum controllable current in the MCT is degraded drastically as the supply voltage is increased.
（MCTの制御可能な最大電流は，供給電圧が増加するにつれて大幅に減少するということが報告されている）

□ When the surface of the dielectric coating becomes wet due to condensation or rain,the reflector antenna performance can be seriously degraded.
（誘電体被覆の表面が結露，あるいは雨によってぬれる時，反射形アンテナの性能は著しく劣化する場合もある）

□ Phase-noise leakage degrades a receiver's performance in several ways.
（位相雑音の漏れは，いくつかの点で受信機の性能を低下させる）

□ The gate resistance R_g does not degrade f_{Tx} and, more interestingly, neither does R_{gs}.
（ゲート抵抗R_gによってf_{Tx}は低下せず，さらに面白いことに，R_{gs}も低下しない）

ポイント☞［類］drop, lower, deteriorate　n:degradation 低下・劣化
☞performance degradation(性能劣化)

demodulate

(demodulated, demodulated, demodulating)

〈他〉(変調波を)復調する

☐ The spread spectrum signals are easily demodulated by the SAW matched-filter.
(スペクトル拡散信号は,SAWマッチドフィルタによって容易に復調できる)

☐ The converted ASK signal was demodulated by a square-law detector.
(変換されたASK信号は,2乗検波器により復調された)

☐ As a result,this 40 MHz signal acts as a carrier that is demodulated in real time with an FM discriminator.
(その結果,この40 MHzの信号は,FM弁別器によって実時間で復調される搬送波としての役割を果す)

☐ Signals near the dither frequency are obtained by demodulating the sensor output using a phase sensitive detector.
(ディザー周波数付近の信号は,位相敏感検波器を用いてセンサの出力を復調することで得られる)

ポイント☞n:demodulation 復調
☞demodulator は復調器.

demonstrate

(demonstrated, demonstrated, demonstrating)

〈他〉(学説・真理などを)実証する,証明する,論証する
demonstrate that節:～ということを証明する

☐ Many A-D converters have already been demonstrated in superconductive technology.
(超伝導技術においては,多くのA/D変換器がすでに実証されている)

☐ The accuracy of the numerical method is demonstrated by comparison with measured values.

（その数値手法の精度を，測定値と比較することで実証する）

☐ We are now ready to demonstrate the application of these results to some problems of interest.
（これで，これらの結果の興味深い問題への応用を実証する準備ができた）

☐ A linear time-invarant system was used as an example to demonstrate the feasibility of the proposed control approach.
（提案された制御法の実現可能性を実証するための一例として，線形時不変システムを用いた）

☐ The results demonstrated that a square-wave power supply was not necessary for a simple Josephson digital circuit.
（方形波電源は，単純なジョセフソン・ディジタル回路には必要ないということを，この結果によって証明した）

ポイント☞［類］verify, validate　　n:demonstration 実証・証明・論証

depend
(depended, depended, depending)
〈自〉 depend on（upon） ～に依存する，～に左右される，～によって決まる

☐ For this restricted case,the backscatter coefficient depends upon frequency.
（この限られた事例の場合，後方散乱係数は周波数に依存する）

☐ The conversion efficiency depends on the characteristic impedance of the resonant circuit.
（その変換効率は共振回路の特性インピーダンスによって決まる）

☐ The number of gates in the OR array depends on the PLA design, but is usually between 5 and 12 gates.
（ORアレイのゲート数はPLAの設計に左右されるが，普通は5～12ゲートである）

ポイント☞［類］rely on　　n:dependence 依存　　adj:dependent 依存の

☞be dependent on(～に依存している)↔be independent of(～から独立している)

☞frequency dependence(周波数依存性)

deposit

(deposited, deposisted, depositing)

〈他〉（薄膜などを）堆積させる，蒸着させる

□ These films have been deposited on silicon substrates without the use of a buffer layer.

（これらの膜を，緩衝層を用いないでシリコン基板上に堆積した）

□ The SiON layer is removed,and a layer of TiWN is deposited by reactive sputtering.

（SiON層を除去してから，TiWN層を反応性スパッタリングによって堆積させる）

□ The niobium films are deposited to a thickness of 300 nm by RF sputtering and patterned by reactive ion etching.

（ニオブ膜を，RFスパッタリングによって300 nmの厚さまで堆積し，反応性イオンエッチングによってパターンを形成する）

□ Over this layer a thin film of resistive material such as Ta_2Al or HfB_2 is sputter-deposited.

（この層一面に，Ta_2AlやHfB_2といった抵抗材料の薄膜をスパッタ蒸着する）

□ To reduce this loss,it is necessary to deposit HTS materials such as YBCO or TBCCO on the ferroelectric substrates.

（この損失を低減するには，YBCOやTBCCOのようなHTS材料を，強誘電基板上に堆積させる必要がある）

ポイント☞n:deposition 堆積

☞名詞としては「堆積物」．chemical vapor deposition(化学蒸着法)

derive

(derived, derived, deriving)

〈他〉 (性質・知識・利益などを)引き出す，導き出す
(物・事の)由来をたずねる
derive A from B：AをBから引き出す

☐ In this paper the dyadic Green's functions for a general uniaxial medium were derived.
(本論文では,一般的な一軸媒質に対するダイアデックグリーン関数を導き出した)

☐ The equations of motion can be derived from Newton's laws or Lagrangian techniques.
(運動方程式はニュートンの法則,あるいはラグランジュ法から導き出すことができる)

☐ We derive a robust stabilization criterion for the nonlinear time-varying multivariable stochastic feedback system shown in Fig.1.
(図1に示す非線形時変多変数確率フィードバックシステムのためのロバスト安定化基準を導き出した)

☐ In Section Ⅲ, we derive several properties for the multistage median filter proposed in [1].
(第3節で，[1]で提案された多段メディアンフィルタの性質を導き出す)

☐ The thickness, derived from C/V curves, is always smaller than the one given by TEM.
(C/V曲線から導き出された厚みは,TEMによって得られた厚みよりも常に小さい)

ポイント☞［類］draw
☞「由来をたずねる」の意味では受動態で用いることが多い.

describe

(described, described, describing)

〈他〉（物・事について）述べる，記述する

☐ This interface is described in more detail in the next section.
（このインタフェースは次節でさらに詳しく述べられる）

☐ Under these conditions, the problem can be described by scalar quantities.
（これらの条件の下で，この問題はスカラー量で記述できる）

☐ This paper describes a novel accelerometer based on electromagnetic induction.
（本論文では，電磁誘導に基づく新しい加速度計について述べている）

☐ This paper describes the outline of the newly developed digital camera system.
（本論文では，新たに開発されたディジタルカメラシステムの概要について述べる）

☐ Cartesian coordinates are used to describe the position and orientation of objects.
（デカルト座標は物体の位置と方位を表すのに用いられる）

☐ The natural frequencies are identical to those described by the singularity expansion method.
（その固有振動数は，特異点展開法によって記述される固有振動数に等しい）

☐ Such models are often used to describe the input-output relationship of practical nonlinear systems.
（実際の非線形システムの入出力関係を記述するために，そのようなモデルがしばしば用いられる）

☐ Based on this relationship, we address the problems described at the end of Section 3.
（この関係に基づいて，第３節の終わりで述べられた問題を取り扱う）

☐ We review some basic concepts in fuzzy sets and systems theory which are

useful in describing the fuzzy systems.
（ファジィシステムを記述するのに役立つファジィ集合とシステム理論の基本概念について概観する）

ポイント☞［類］state　n:description 記述　adj:descriptive 記述的な
☞「物・事の特徴や状況について詳細に述べる」のニュアンスがある．

design
(designed, designed, designing)
〈他〉（物を）設計する

□ An I/O robust controller will be designed for the robot using the nonlinear mapping (16).
（非線形写像(16)を用いて，ロボット用にＩ／Ｏロバスト制御装置を設計する）

□ Different types of sensors have been designed for factory automation, measurement and safety.
（さまざまなタイプのセンサが，工場自動化，測定,安全性のために設計されてきた）

□ We designed CMOS operational amplifiers by recycling previously designed layouts.
（前に設計されたレイアウトを再利用することで,CMOS演算増幅器を設計した）

□ This concept can be extended to design a multiinput, multioutput gate system.
（多入力，多出力のゲートシステムを設計するのに，この概念は拡張できる）

□ In designing the packaging of an optical device, all of the factors which affect conventional electronic devices need to be considered.
（光学デバイスのパッケージングを設計する際,従来の電子デバイスに影響を及ぼす要因すべてを考慮する必要がある）

☐ These bounds are required in designing robust controls.
（これらの限界は，ロバスト制御装置を設計する際に必要となる）

ポイント☞名詞としては「設計・デザイン」.

☞automatic design(自動設計)，logic design(論理設計), software design (ソフトウェア設計)，system design(システム設計)

detect

(detected, detected, detecting)

〈他〉（信号・欠陥・化学物質などを）検出する

☐ The transmitted light is detected either by an infrared camera system or by a germanium photodiode.
（透過光は赤外線カメラシステム，あるいはゲルマニウム・フォトダイオードによって検出される）

☐ Acoustic emission testing is a well-known test technique that detects the pressure wave generated during an electrical discharge.
（アコースティックエミッション試験は，電気放電中に発生する圧力波を検出する有名な手法である）

☐ The SAW FM discriminator can detect the received DPSK signal.
（SAW FM弁別器は受信されたDPSK信号を検出できる）

☐ A phototransistor was employed to detect the reflected beams and convert them to an electrical signal.
（反射ビームを検出し，それを電気信号に変換するのに，フォトトランジスタを使用した）

☐ Auger electron spectroscopy was not able to detect a chemical interaction at the $WSi_{0.4}$/GaAs interface.
（オージェ電子分光法では，$WSi_{0.4}$/GaAs界面での化学的相互作用は検出できなかった）

☐ The circuits used to detect the phase are simple and inexpensive.
(位相を検出するために用いられた回路は単純で安価である)

☐ Of particular interest is the ability to detect defects in an interface between two materials.
(とりわけ興味深いことは,二つの材料間の界面内の欠陥を検出する能力である)

☐ Ultrasonic inspection is an established non-destructive evaluation method for detecting defects in structural components.
(超音波検査は,構造要素の欠陥を検出するための確立した非破壊評価法である)

ポイント☞n:detection 検出　　adj:detectable 検出できる
☞detector は検出器．signal detection theory(信号検出理論)，signal detector(信号検出器)

determine
(determined, determined, determining)
〈他〉（物・事を）決定する
（位置・形状などを）測定する

☐ In the quasi-TEM theory, the propagation constants are determined from the transmission line circuit parameters.
(準TEM理論の場合,伝搬定数は伝送線路の回路パラメータから決定される)

☐ The response of a matched filter is determined by this signal model.
(整合フィルタの応答は，この信号モデルで決まる)

☐ The film thickness can be accurately determined from data obtained in the 10-75° range of incidence angle.
(膜厚は10～75°の範囲の入射角に対して得られたデータから正確に計算できる)

☐ The magnetic properties were determined in static magnetic fields and in alternating magnetic fields of different frequencies.
（磁気的性質は静磁場中と，周波数が異なる交番磁場中で測定された）

☐ No optical filtering was used between the fibre amplifier and receiver, so the spontaneous noise bandwidth was determined by the gain of the amplifier.
（光学フィルタリングはファイバ増幅器と受信機の間で使用されなかった．それで，自然雑音の帯域幅は増幅器の利得によって求めた）

☐ In designing a filter, the chosen damping coefficient will determine the shape of the filter response.
（フィルタを設計する際，選ばれた減衰係数はフィルタ応答の形状を決定する）

ポイント☞［類］decide, settle　　n:determination 決定
☞「考慮の結果，はっきりと決定する」のニュアンスがある．

develop
(developed, developed, developing)
〈他〉（物を）開発する

☐ Two software packages have been developed to operate the system.
（そのシステムを作動させるために，二つのソフトウェアパッケージが開発された）

☐ An all thin-film SC disk resonator is currently being developed.
（全薄膜SCディスク共振器が現在開発中である）

☐ We are developing a superconducting linear induction motor for steelmaking processes using ultra-fine filamentary NbTi wires.
（超微細線状NbTi線を用いて，製鋼工程用の超伝導リニア誘導電動機を開発している）

☐ It is possible to develop an algorithm for estimating cloud attenuation as a

function of these parameters.
(これらのパラメータの関数としての雲減衰を推定するためのアルゴリズムを開発することは可能である)

□ It is important to develop plasma diagnostic methods suitable for these applications.
(これらの用途に適したプラズマ診断法を開発することは重要である)

□ In this case, various techniques developed for linear systems can be used to realize the linear subsystems and to determine their system order.
(この場合，線形システム用に開発されたさまざまな手法が，線形サブシステムを実現し，そのシステムの次数を決定するのに使うことができる)

ポイント☞［類］exploit　　n:development 開発
☞software development(ソフトウェア開発)

differ
(differed, differed, differing)
〈自〉（物・事が）異なる
differ from A：Aと異なる
differ in A：Aの点で異なる

□ This paper differs from earlier analyses of the data quantization problem in four respects.
(本論文は，データ量子化問題に対する従来の解析とは，四つの点で異なる)

□ Optimal control differs from other forms of control in that the desired values of the controlled variables are unknown.
(最適制御は，制御される変数の目標値は未知であるという点で他の形式の制御とは異なる)

□ The various detectors also differ in complexity.
(これら種々の検出器は複雑さの点でも異なる)

☐ Electric and magnetic fields differ in their generation, coupling mechanism and mitigation procedures.
（電場と磁場はそれらの発生，結合機構，緩和手順の点で異なる）

ポイント☞［反］agree(一致する)　［類］vary　n:difference 相違
adj:different 異なった
☞differ from＝be different from

diffuse
(diffused, diffused, diffusing)
〈他〉（ガス・液体・熱・光などを）拡散する
〈自〉（ガス・液体・熱・光などが）拡散する

☐ The 675 Å-thick titanium pattern was diffused into the substrate for 6 h at T＝1,050℃.
（675Å厚のチタンパターンはT＝1,050℃で6時間，基板内で拡散した）

☐ Superconducting pairs of electrons can diffuse into the doped simiconductor and make it weakly superconductive.
（超伝導電子対がドープされた半導体内に拡散し，その半導体を弱い超伝導状態にする）

☐ The hot electrons diffuse into the fins, where they rapidly transfer energy to other electrons.
（熱い電子はフィン中に拡散するが，ここで急速にエネルギーを他の電子に移す）

ポイント☞n:diffusion 拡散
☞impurity diffusion(不純物拡散)

digitize
(digitized, digitized, digitizing)
〈他〉（信号などを）ディジタル化する，計数化する

☐ Signals were digitized at 250 MHz.

(信号は250 MHzでディジタル化された)

☐ As we are interested in the magnitude of the signal, the signal is then rectified and finally digitized.
(信号の大きさに関心をもっているので，次に信号を整流し，最後にディジタル化する)

☐ The signals were digitized by a software-controlled transient recorder at asampling rate of 40 MHz.
(40 MHzの標本化周波数において，ソフトウェア制御された過渡記録計によって信号はディジタル化された)

ポイント☞n:digitization ディジタル化　adj:digital ディジタルの
☞digital↔analog(アナログの)．digital-to-analog converter(ディジタル－アナログ変換器)，digital computer(ディジタル計算機)，digital recording(ディジタル記録)

discuss
(discussed, discussed, discussing)
〈他〉(物・事について)論じる，議論する

☐ The modeling of this antenna has been discussed in detail by King.
(このアンテナのモデリングについては，キングによって詳細に論じられた)

☐ This condition is discussed in more detail in Section V.
(この条件は，第5節でより詳細に論じられる)

☐ We discuss continuous-time systems, but the results are applicable to discrete-time systems as well.
(連続時間システムについて議論するが，この結果は離散時間システムにも適用できる)

☐ We discuss why improvement of data error rate arises.
(データの誤り率の向上がなぜ起こるのかということについて論じる)

☐ In Sections 4 and 5, we shall discuss how to obtain numerical solutions of (8).
（第4節と5節で，(8)の数値解をどのようにして得るかという方法について論じる）

☐ To discuss the basic relationships, it is enough to consider at present only a cubic crystal.
（この基本的関係を論じるためには，今のところ,立方晶系結晶のみを考察すれば十分である）

ポイント☞［類］argue, dispute　　n:discussion 討論
☞「あらゆる角度から問題解決のために論じる」のニュアンスがある．他動詞なのでdiscuss about は誤り．

display
(displayed, displayed, displaying)
〈他〉（文字・図形などを)表示する

☐ The time recorded by the analyzer clock is displayed on the PC screen.
（アナライザ・クロックによって記録された時間はPC画面に表示される）

☐ In the CT scan system, images are displayed on the video screen in a 512×512 pixels.
（そのCT走査システムでは，画像は512×512画素のビデオスクリーンに表示される）

☐ The system displays reflected signals on an oscilloscope.
（システムは反射信号をオシロスコープに表示する）

ポイント☞名詞としては「表示(装置)・ディスプレイ」.
☞color display(カラーディスプレイ), graphic display(グラフィックディスプレイ), plasma display(プラズマディスプレイ)

distinguish

(distinguished, distinguished, distinguishing)

〈他〉（物・事を）区別する，識別する

distinguish A from B：AをBと区別する

〈自〉（物・事の間の相違を）見分ける，区別する

distinguish between A and B：AとBを見分ける

□ These offsets can be distinguished from the backscatter noise by comparatively long fluctuation period.

（比較的長い変動周期によって，これらのオフセットは後方散乱雑音と区別することができる）

□ The receiver can distinguish between the symbol value and the intersimbol inteference.

（その受信機は，符号値と符号間干渉を区別することができる）

□ Under certain conditins, (14) can be used to distinguish the LNL model from the $PLNL_2$ model.

（一定の条件の下で，(14)はLNLモデルを$PLNL_2$モデルと区別するのに用いることができる）

ポイント☞［類］differentiate, discriminate　n:distinction 区別　adj:distinct 異なった

☞「物・事の特徴・特性を見分けて区別する」のニュアンスがある．

distort

(distorted, distroted, distorting)

〈他〉（物体・信号・波形を）歪ませる，変形させる

□ This technique may produce a very large phase error when an input wave is distorted by a harmonic.

（入力波が高調波によって歪む時，この手法は非常に大きな位相誤差を生み出すことがある）

☐ If the input wave is distorted by the fundamental or odd harmonics, the method presented in [12] becomes not applicable.
（入力波が基本波，あるいは奇数調波によって歪むならば，[12]で提出された方法は適用できなくなる）

☐ A median filter can eliminate impulsive noises, however it distors the signal waveform.
（メディアンフィルタはインパルスノイズを除去できるが，信号波形を歪ませる）

☐ This paper presents the error analysis of phase measurement of distorted waves using existing phasemeters.
（本論文では，既存の位相計を用いて，歪み波の位相測定の誤差解析を示す）

ポイント☞n:distortion 歪・変形
☞harmonic distortion(高調波歪), frequency distortion(周波数歪), phase distortion(位相歪)

divide

(divided, divided, dividing)

〈他〉（物・事を）分割する，分ける
divide A into B：AをBに分割する

☐ Optically controlled semiconductor optical signal processing devices can roughly be divided into two separate groups.
（光学的に制御された半導体光信号処理デバイスは，大きく二つの異なったグループに分けることができる）

☐ The hardware for the harmonic analyzer can be divided into a digital and an analog section.
（調波分析器のためのハードウェアは，ディジタル部とアナログ部に分割できる）

☐ The input coupler divides the input power equally into two single-ended

amplifiers.
(その入力結合器は入力電力を二つのシングルエンド増幅器に均等に分割する)

□ The diode model, which is shown in Fig.1, divides the diode into N constant temperature regions.
(図1に示すダイオードモデルでは，ダイオードはN個の一定温度の領域に分割されている)

□ The most widely used concept is to divide a ciucuit into a number of uniform sections in the direction of propagation.
(最も広く用いられる概念は，回路を伝搬方向に多くの均一な部分に分割することである)

□ Dividing the power loss into hysteresis loss and eddy current loss, the frequency dependence of the eddy current was found to vary with the magnitude of the dc resistivity.
(電力損失をヒステリシス損と渦電流損に分けると，渦電流の周波数依存性は直流抵抗率の大きさによって変わるということがわかった)

ポイント☞［類］split　　n:division 分割
☞「基準・計画・寸法などに従って注意深く分割する」のニュアンスがある．divider はデバイダ・分周器．

do

(did, done, doing)

〈他〉（実験・解析などを）行う，する

□ A statistical analysis was done on all 210 sets of cloud data.
(210組の雲データすべてに対して統計解析を行った)

□ The testing was again done using the 4-point method.
(4点法を用いて試験を再び行った)

□ The above analyses were done under adiabatic boundary conditions.
（上述の解析を断熱境界条件の下で行った）

□ Stability analysis and control design will be done using Lyapunov's direct method.
（安定解析と制御設計を，リャプノフ直接法を用いて行う）

□ For example, CD-ROM applications require simple decoders,but the encoding process does not have to be done in real time.
（例えば，CD-ROMの用途には簡単なデコーダが必要だが，符号化処理は実時間で行う必要はない）

□ We have to do final tests on the motor.
（電動機の最終検査をしなければならない）

ポイント☞［類］make, conduct, perform, run
☞本動詞の他に助動詞や代動詞としても用いる．do away with(～をやめる), do nothing but(～ばかりする), do without(～なしですます)

draw
(drew, drawn, drawing)
〈他〉（結論などを）導く，得る
（線などを）引く，描く

□ However, very useful conclusions can be drawn from the data presented here.
（しかしながら，ここで提示したデータから非常に有用な結論を導き出すことができる）

□ This circuit can be drawn in a signal flow graph.
（この回路はシグナルフローグラフで描くことができる）

□ There are several general conclusions we can draw from Figs.1 and 2.
（図１と２から導き出せる一般的結論がいくつかある）

☐ We can draw two conclusions from Fig.3.
(図 3 から二つの結論が得られる)

☐ In the design process,the designer can draw layout easily without considering complicated design rule constraints.
(その設計過程では，設計規則の複雑な制約を考慮することなく，設計者は容易にレイアウトを描くことができる)

ポイント☞［類］derive
☞drawer は引出し・製図工，drawing は製図. drawing board(製図板)

drive
(drove, driven, driving)
〈他〉（回路・機械などを）駆動する

☐ The four MOSFETs were driven directly by a driver IC.
(4 個のMOSFETはドライバICで直接駆動される)

☐ The joints of the manipulator are driven by actuators.
(マニピュレータの関節はアクチュエータで駆動される)

☐ This periodic motion drives a rotor intermittently.
(この周期運動は回転子を断続的に駆動する)

☐ The differential output voltage of the amplifiers can drive the differential flash ADC directly.
(増幅器の差動出力電圧は差動フラッシュADCを直接駆動できる)

☐ This photoisolator can drive a 1.2 A class load directly.
(このフォトアイソレータは，1.2アンペアクラスの負荷を直接駆動することが可能である)

☐ The goal of this project is to achieve complete intergration of the scanning circuitry necessary to drive an active-matrix LCD.

（このプロジェクトの目標は，アクティブマトリックスLCDを駆動するのに必要な掃引回路を完全集積化することである）

□ The Ward Leonard system is employed to drive the dc motor.
（ワードレオナード方式は直流電動機を駆動するのに用いられる）

□ This paper provides a general relation for an energy stored in a reactive network driven by a sinusoidal source.
（本論文では，正弦波源によって駆動される無効回路網に蓄積されるエネルギーに対する一般的関係を与える）

ポイント☞名詞としては「駆動」だが，可算名詞として「駆動装置」としても使われる．

☞disk drive(ディスク駆動機構)，tape drive(テープ駆動機構)，·drive motor(ドライブモータ)

E

eliminate

(eliminated, eliminated, eliminating)

〈他〉（物・事を)除く，除去する

□ The errors of the sensors are eliminated by preprocessing the raw data.
（センサのエラーは生データを前処理することで除去される）

□ The high-frequency noise can be eliminated by using the Butterworth low-pass filter wtih 1.5 Hz cutoff frequency.
（その高周波雑音は，1.5 Hzの遮断周波数を有するバターワース低域フィルタを用いることで除去できる）

□ These techniques provide high resolution and eliminate the feedback loop

instability problem.
(これらの手法は高分解能をもたらし, フィードバックループ不安定性問題をなくする)

☐ The model does not eliminate the necessity of measuring the product.
(そのモデルでは, 製品を測定する必要性がなくならない)

☐ Some methods were developed which effectively eliminated this problem for systems having time delay of less than 100 ms.
(時間遅れが100 ms以下のシステムのために, この問題を効果的に取り除く方法が開発された)

☐ Transform coding techniques are used to eliminate correlation between image pixels.
(変換符号化法は, 画素間の相関を除くために用いられる)

ポイント☞ ［類］remove, get rid of　　n:elimination 除去
☞「ある手順に従って望ましくない物を除く」のニュアンスがある.

emit
(emitted, emitted, emitting)
〈他〉（熱・光・音・ガスなどを）放つ，放出する，放射する

☐ A light source emits rays in all directions to generate an elementary spherical wave,each ray propagating along straight lines.
(光源は四方に光線を発し, 基本球面波を発生させ, 各光線は直線に沿って伝搬する)

☐ In this model the transmitter emit either an infinite wave train of a single frequency or a pulse of finite length.
(このモデルでは, 送信機は単一周波数の無限波列, あるいは有限長のパルスを放射する)

☐ The phase conjugation of the optical beam emitted by semiconductor lasers is

usually achieved by degenerate four-wave mixing in photorefractive media.
(半導体レーザから放射された光ビームの位相共役は，通常は光屈折媒質の縮退四光波混合によって達成される)

☐ Only this small portion of the luminous energy emitted by the lamp reaches the screen.
(ランプから放射されたルミナスエネルギーのうち，このほんの一部分だけがスクリーンに達する)

ポイント☞［類］radiate
☞放出・放射は emission. photoelectron emission(光電子放出)

employ
(employed, employed, employing)
〈他〉（物・手段などを)使用する，用いる，利用する

☐ This amplifier is employed extensively in many areas of applied electronics.
(この増幅器は，応用電子工学の多くの分野で広範に用いられている)

☐ As a result, these local Lyapunov functions cannot be directly employed in direct methods for power system transient stability analysis.
(その結果，これらの局所リャプノフ関数は，電力系統の過渡安定解析のための直接法には直接使用できない)

☐ Today many coherent optical transmission systems employ frequency shift keying (FSK).
(今日では，コヒーレント光伝送方式の多くは周波数偏移変調（FSK)を用いている)

☐ The neural network employed here is a multilayered feedforward feedback type network, the structure of which is shown in Fig.1.
(ここで利用されるニューラルネットワークは多層フィードフォワード・フィードバック型ネットワークであり，その構造は図１に示している)

ポイント☞［類］use, utilize, exploit, make use of, take advantage of n:employment 使用・利用

enable

(enabled, enabled, enabling)

〈他〉（事を）可能にする

enable A to do：Aに～することを可能にさせる

□ The use of these techniques enable sensors to be constructed rapidly and economically.
（これらの手法を使うことで，センサを速く，かつ経済的に組み立てられるようになる）

□ This technique enables very good laser diode performance and reliability to be obtained.
（この手法によって，レーザダイオードの非常に優れた性能と信頼性を得ることができるようになる）

□ We propose a video coding scheme that enables hierarchical video signal processing.
（階層ビデオ信号処理を可能にするビデオ符号化方式を提案する）

□ This feature enables high-accuracy recording of temporal Fourier components of the signal light.
（この特徴により，信号燈の時間フーリエ成分の高精度記録が可能になる）

ポイント☞受動態は不可．主として無生物主語の構文に用いられる．

enhance

(enhanced, enhanced, enhancing)

〈他〉（価値・性能などを）高める，増大する

□ The signal-to-noise ratio is enhanced by using a relatively high phase dither frequency.
（そのS/N比は，比較的高い位相ディザー周波数を用いることで向上する）

☐ A bus architecture generally enhances testability of the circuit.
（バスアーキテクチャは一般に，回路の試験容易性を高める）

☐ An objective of the design is to enhance gain and bandwidth performance by preventing parasitic oscillation.
（この設計の目的は，寄生振動を防ぐことで利得と帯域幅の性能を向上させることである）

☐ In order to enhance the performance of microwave spectrometers,a variety of modulation techniques have been employed such as Stark modulation and Zeeman modulation.
（マイクロ波分光計の性能を高めるために，シュタルク変調，ゼーマン変調といったさまざまな変調手法が用いられてきた）

ポイント☞［類］increase, rise, improve, upgrade　n:enhancement 向上・増進
☞「望ましいことや好ましいことを高める」のニュアンスがある．
image enhancement(画像強調)

ensure

(ensured, ensured, ensuring)

〈他〉（事を）保証する，確実にする
ensure that節：～ということを保証する，確実にする

☐ It is well known that optimal logic synthesis can ensure fully testable combinational logic designs.
（最適論理合成は，完全に試験可能な組合わせ論理設計を保証できるということは，よく知られている）

☐ We design adaptive laws to update the controller parameters, which ensure closed-loop signal boundedness.
（制御装置のパラメータを更新するための適応則を設計するが，この適応則は閉ループ信号の有界性を保証する）

☐ This choice of the sign of the magnetic current ensures that the tangential

magnetic field is continuous across the aperture.
(磁流の符号をこちらに選べば，接線方向の磁場は開口全体にわたって連続的であるということが確実となる)

□ The location of these zeros is of great importance to ensure an accurate and exact evaluation of (1).
(これらの零点の位置は，(1)の正確で厳密な評価を確実なものにするためには非常に重要である)

□ Eight bits are required to ensure a high fidelity image.
(高忠実度画像を保証するには８ビットが必要である)

ポイント☞［類］assure, insure, gurantee, warrant
☞類語の中では ensure の使用頻度が高い.

equal
(equaled, equaled, equaling)
〈他〉（数量・大きさなどの点で）等しい

□ The voltage between the base and the anode equals V_{bc} because of assumption (*f*).
(ベース・陽極間の電圧は仮定(*f*)のためにV_{bc}に等しい)

□ The time variation of the radiated waves approximately equals that of a Gaussian pulse.
(その放射波の時間変動は，ガウスパルスの時間変動にほぼ等しい)

□ The counter output is a 16-bit word which equals the number of clock pulses in one input period.
(カウンタの出力は，1 入力周期中のクロックパルスの数に等しい16ビットワードである)

□ The total processing time T_t equals the transmission time plus the computing time of the processor.

（全処理時間T_tは伝送時間にプロセッサの計算時間を足したものに等しい）

□ This relation is obtained from the fact that the magnetic field on the surface nearly equals Hy/cos Θ.
（この関係は，表面上の磁場はHy/cos Θにほぼ等しいという事実から得られる）

ポイント☞n:equality 等しいこと・等式
☞進行形は不可．形容詞としては「等しい」．形容詞用法では be equal to（～に等しい）の形になる．

equip

(equipped, equipped, equipping)

〈他〉（物を）備えつける，装備する
equip A with B：AにBを備える

□ The shielding configuration was eqipped with Type N connectors.
（その遮蔽構成には，N型コネクタが装備された）

□ The shaft of the DC servomotor is equipped with a potentiometer in addition to an optical attenuation element.
（直流サーボモータのシャフトには，光減衰要素に加えて電位差計が備えつけられている）

□ These reflectometers are equipped with broad-band light sources.
（これらの反射率計には，広帯域光源が備えつけられている）

ポイント☞［類］furnish　n:equipment 装備・装置
☞「仕事・業務などに必要な機器などを備え付ける」のニュアンス．

establish

(established, established, establishing)

〈他〉（理論・学説などを）立証する，証明する
（値などを）確定する，決める

（規則・制度などを）確立する

☐ Liquid-crystal devices (LCD's) are now well-established as the preferred displays in many applications,including watches,calculators, and lap-top computers.
（今では液晶デバイス(LCD)は，時計，計算器，ラップトップコンピュータを含む多くの用途で，好ましいディスプレイとして十分に立証されている）

☐ The procedure establishes a necessary and sufficient condition for the assessment of multivariable system stability.
（その手順によって，多変数システムの安定性を評価するための必要十分条件は確立される）

☐ In practice, it is very difficult to establish precisely the value of this coefficient.
（実際には，この係数の値を正確に決めることは非常に困難である）

☐ Nunerical examples are treated to establish the validity of the proposal.
（そのプロポーザルの妥当性を確立するために，数値例を扱う）

☐ The following lemma is useful for establishing exponential convergence for a class of time-varying systems.
（次の補助定理は，ある種の時変システムのための指数収束を確定するのに役立つ）

ポイント☞n:establishment 設立・確立・立証

estimate

(estimated, estimated, estimating)

〈他〉（能力・特性などを）評価する，推定する

estimate A at (to be) B：AをBであると推定する

☐ Uncertainty of the temperature measurements was estimated to be ±0.3℃.
（温度測定値のあいまいさは，±0.3℃と推定された）

☐ The parastic capacitance between the input and output terminals due to wirings was estimated to be 0.08 pF.
(配線による入出力端子間の寄生容量は，0.08 pFと推定された)

☐ The accuracy of the proposed method has been estimated by computer simulations.
(提案された方法の精度は,コンピュータシミュレーションによって評価された)

☐ This algorithm will estimate the system cut-off frequency and gain on-line.
(このアルゴリズムはシステムの遮断周波数と利得をオンラインで推定できる)

☐ In this case a Kalman filter might be used to estimate the states of the aircraft, based on radar measurements.
(この場合，カルマンフィルタはレーダによる測定値に基づいて,航空機の状態を推定するのに使用できるかもしれない)

☐ In this section, we will present algorithms for estimating parameters of the PLN model.
(この節では，PLNモデルのパラメータを推定するためのアルゴリズムを示す)

☐ Experimental results for water and methanol are compared with estimated values.
(水とメタノールに対する実験結果を推定値と比較する)

ポイント☞［類］assess,evaluate,value,appreciate　n:estimation 評価・推定
☞「数量・価値などを主観的に見積もる」のニュアンスがある．名詞としては「評価・推定値」.

evaluate

(evalutated, evaluated, evaluating)

〈他〉（価値・能力などを）評価する
（数式などの）値を求める

☐ Bending losses of several fibers were evaluated by the proposed method.
（いくつかのファイバの曲げ損失は提案された方法で評価された）

☐ The accuracy and limitations of the method have been evaluated via computer simulations.
（その方法の精度と限界は，コンピュータシミュレーションによって評価された）

☐ It is evident that the required integrals can always be evaluated using numerical intergration.
（必要とされる積分は，常に数値積分法を用いて値を求めることができるということは明らかである）

☐ Computer simulations was performed to evaluate the performance of the proposed algorithm.
（提案されたアルゴリズムの性能を評価するために，コンピュータシミュレーションを実行した）

☐ Several measurements were made to evaluate the equipment performance.
（装置の性能を評価するために測定を行った）

☐ In order to evaluate the corrosion resistance of the Co-Ni-Fe-Pd film, we investigated the change of magnetic flux of the film.
（Co-Ni-Fe-Pd膜の耐食性を評価するために，その膜の磁束の変化を調べた）

ポイント☞［類］assess, estimate, value,appreciate　n:evaluation 評価
☞「物の金銭以外の価値・重要性を判断する」のニュアンスがある.

evaporate

(evaporated, evaporated, evaporating)

〈他〉（薄膜・層などを）蒸着させる

☐ A 100 nm-thick layer of metal is evaporated normal to the sample surface.
(100 nm厚の金属層をサンプル表面に垂直に蒸着する)

☐ These films can be evaporated on a substrate.
(これらの膜は基板上に蒸着できる)

☐ Ohmic contacts were established by evaporating a layer of Al onto the substrate.
(Al層を基板上に蒸着させることで，オーム接点を作った)

ポイント☞n:evaporation 蒸着
☞vacuum evaporation(真空蒸着)

examine
(examined, examined, examining)
〈他〉（物・事を）検査する，検討する，調査する

☐ These samples were examined using scanning electron microscopy.
(これらのサンプルは走査電子顕微鏡法を用いて検査された)

☐ This paper examines some aspects of high level computer vision systems.
(本論文では，高レベル・コンピュータビジョンシステムのいくつかの側面について検討する)

☐ It is possible to examine the harmonic distortion of the power line voltage waveform.
(電力線の電圧波形の高調波歪みを調べることは可能である)

☐ Microstrip HTS phase shifters can be analyzed by separately examining the two components of the ferroelectric phase-shifter loss:dielectric loss and conductive loss.
(マイクロストリップHTS移相器は，強誘電体移相器の損失の二つの成分，つまり誘電損と導電損を別々に調べることで解析できる)

☐ Before examining VLSI architectures, it is useful to review these image

compression techniques.
(VLSIアーキテクチャを検査する前に，これらの画像圧縮手法を概観することは有用である)

ポイント☞[類]check, investigate, inspect, discuss, study, explore　n:examination 検査・検討・調査
☞「性質・状態を知るために詳しく調べる」のニュアンスがある.

excite
(excited, excited, exciting)
〈他〉（電磁石・コイルなどを）励磁する
（アンテナなどを）励振する

□ One core is excited by 2 and 3 planar coils and the other by 4 and 5.
(一方のコアは２と３の平面コイルで，他方のコアは４と５の平面コイルで励磁される)

□ The transformer was excited by a constant-current power supply.
(変圧器は定電流電源により励磁された)

□ The antenna is excited for linear polarization along the y axis.
(アンテナはy軸に沿って直線偏波するように励振される)

□ The applied FM signal will excite the two collinear waves which are propagating in the same direction.
(加えられたFM信号は，同じ方向に伝搬している二つの共線波を励振する)

ポイント☞n:excitation 励磁・励振
☞exciting current(励磁電流)，exciting pulse(励振パルス)

execute
(executed, executed, executing)
〈他〉（計画などを）実行する，実施する，遂行する

□ The fault diagnosis is executed from a series of individual tests of various

parts of the circuitry.
(回路のさまざまな部分を一連の個別試験を行うことで故障診断を実施する)

□ The microcomputer executes the speed control of this motor.
(マイクロコンピュータはこの電動機の速度制御を実行する)

□ This technique provide a very simple way to execute parallel optical logic operations.
(この手法は，並列光論理演算を実行するための非常に簡単な方法を提供する)

ポイント☞［類］carry out, implement, run n:execution 実行 adj:executive 実行する
☞「目標・計画・命令などを実行し，やり遂げる」のニュアンスがある.

exhibit
(exhibited, exhibited, exhibiting)
〈他〉（特性などを）示す

□ At low frequencies,a linear dependence is exhibited between voltage and frequency, so that the response of an ideal recording head is realized.
(低周波では，電圧と周波数の間に線形依存性が示され，そのため理想的な記録ヘッドの応答が実現する)

□ Sapphire exhibits properties very similar to alumina.
(サファイアはアルミナに非常に似た特性を示す)

□ The magnetization curves of $Co_{30}Ag_{70}$ exhibit hysteresis at room temperature.
(その$Co_{30}Ag_{70}$の磁化曲線は，室温でヒステリシスを示している)

□ Mn-Zn ferrites, usually used as core materials in the switching power supplies, exhibit a high power loss at high frequencies.
(スイッチング電源の鉄心材料に通常用いられるMn-Znフェライトは，高周波

で大きな電力損失を示す）

☐ Lead-zirconate-titanate (PZT5A) is a ferroelectric material that exhibits both pyro and piezoelectric properties.
（ジルコン酸チタン酸鉛（PZT5A）は，焦電気特性と圧電特性の両方を示す強誘電材料である）

ポイント☞［類］show, display, reveal, demonstrate, indicate, illustrate
☞「特徴・問題点などを人の注意を引くように示す」のニュアンスがある．名詞としては「提示・展示」．

exist
(existed, existed, existing)
〈自〉（物が）存在する，実在する

☐ A variation of ±1 Hz existed over a 10-s measurement.
（10秒間の測定にわたって±1 Hzの変化量が存在した）

☐ Very strong noise signals exist in the impulse-voltage measuring environment.
（非常に強い雑音信号が，インパルス電圧の測定環境中に存在する）

☐ In this study, the distribution of cracks exists in an interface between the same material.
（この研究において，亀裂の分布は同じ材料間の界面中に存在する）

ポイント☞n:existence 存在
☞命令形・進行形は不可．in existence(現存の)

expect
(expected, expected, expecting)
〈他〉（物・事を）期待する，予想する，予期する
expect A to do：Aが～することを予期する，期待する
expect that節：～だと期待する，思う，予想する

☐ CDMA is expected to support future land mobile communication system.

(CDMAは，将来の陸上移動通信方式を支援するものと期待されている)

☐ It can be expected that this goal will be attained in the near future.
(この目標は近い将来に達成されるものと期待できる)

☐ As beryllium oxide (BeO) is known as a chemically stable material, Be-doped films are expected to be inactive against acids after the doped beryllium reacts with oxygen.
(酸化ベリリウム (BeO) は化学的に安定な材料として知られているので，Beドープ膜は，ドープされたベリリウムが酸素と反応した後に，酸に対して不活性となることが予測される)

☐ This approximation is expected to result in satisfactory accuracy in the present frequency-range.
(この近似が，現在の周波数範囲において申し分ない精度をもたらすものと期待される)

☐ If v and μ are independent of gate length, we expect a linear relationship between V_L and L_M.
(vとμがゲートの長さに無関係ならば，V_LとL_Mの間に線形関係が予測される)

☐ We expect that such oscillations will eventually disappear.
(そのような発振は結局は消滅するだろうと思われる)

ポイント☞［類］anticipate, predict, forecast　　n:expectation 期待・予想・予期
☞「十分な理由・根拠があり，当然のこととして期待する」のニュアンスがある．

exploit
(exploited, exploited, exploiting)
〈他〉（特性などを）活用する，利用する

□ These signals can be exploited for sophisticated machining control for various purposes including machining quality, safety, maintenance, etc.
(機械加工品質，安全性,保守などを含むさまざまな目的に対する精密な機械加工制御に，これらの信号が利用できる)

□ Several researchers have demonstrated how a piecewise-linear approximation to nonlinear device characterstics can be exploited in very efficient solution schemes.
(非線形デバイス特性に対する区分的線形近似は,非常に効率的な解法でどのように使用できるかということを，数人の研究者が説明している)

□ The IF amplifier can exploit the advantages of semiconductor technology.
(そのIF増幅器は，半導体技術の利点を利用することができる)

□ InP/InGaAs HBT's which exploit the superior transport properties of InGaAs have shown a high cutoff frequency at low current.
(InGaAsのもつ優れた輸送特性を利用したInP/InGaAs HBTは，低電流で高遮断周波数を示した)

□ The design technique has been developed by exploiting the fact that zero-phase FIR filters inherently have centro-symmetry in their frequency responses.
(その設計手法は,零位相FIRフィルタがその周波数応答に中心対称を本質的にもつという事実を利用することで開発された)

□ The technique is based on exploiting the structure of the signals received by an antenna array in both the temporal and spatial frequency domains.
(その手法は，時間および空間周波数領域の両方で,アンテナアレイで受信された信号の構造を利用することに基づいている)

ポイント☞［類］use, utilize, exploy, make use of, take advantage of

explore

(explored, explored, exploring)

〈他〉（可能性などを）調べる，探る，調査する

☐ Recently, barium ferrite particulate media have been explored for use in high-density magnetic recording.
（最近，高密度磁気記録に使用するためのバリウムフェライト粒子媒質が調べられた）

☐ These symmetry properties can be further explored to the complex FIR all-pass filters.
（複素FIR全域通過フィルタに対して，これらの対称性をさらに調べることができる）

☐ In particular, we explore several aspects of this problem.
（特に，この問題のいくつかの側面について探る）

☐ This paper explores the application of recent nonlinear control and observer theory to the practical problem of temperature control in a gas fired furnace.
（本論文では，ガス燃焼炉における温度制御の実際の問題に対する最近の非線形制御とオブザーバ理論の応用について調査する）

☐ Computer graphics provides an excellent mechanism for exploring the colour reproduction characteristics of typical display devices.
（コンピュータグラフィックスは，優れた機構を代表的な表示装置の色再現特性の調査に提供する）

ポイント ☞ ［類］ examine, check, investigate, inspect, discuss, study
n:exploration 探査・調査

expose

(exposed, exposed, exposing)

〈他〉（光・放射線などに）さらす
expose A to B：AをBにさらす

☐ The ceramic cylinder acts as an insulator so that only the electrode surface is exposed to the plasma.
（そのセラミック円筒は，電極表面のみがプラズマにさらされるように絶縁体

の働きをする）

☐ Figure 5 shows the surface of the ceramic sensor after being exposed to the shockwaves.
（図５は，衝撃波にさらされた後のセラミックセンサの表面を示している）

☐ The materials may progressively deteriorate when exposed to light.
（その材料は露光されると次第に劣化する）

ポイント☞n:exposure 暴露・露出
☞exposure meter(露出計)

express
(expressed, expressed, expressing)
〈他〉（事を）表現する，示す，表す

☐ The spectral distribution of blackbody radiation is expressed by Planck's distribution law.
（黒体放射のスペクトル分布は，プランク分布法則で表現される）

☐ The drain-source voltages of the power MOSFETs are expressed by Eqs.(1) and (2), and the resonant frequency of the flyback voltage is expressed by Eq.(3).
（パワーMOSFETのドレインーソース電圧は，方程式(1)と(2)で表され，フライバック電圧の共振周波数は方程式(3)で表される）

☐ If the core is straight and infinitely long,the formulas are expressed in terms of convergent integrals that may be evaluated numerically.
（もしコアが直線で無限長ならば,これらの公式は数値で求められる収束積分によって表される）

☐ We show that the upper bound can be expressed exactly in terms of a modified Bessel function.
（上限は変形ベッセル関数によって厳密に表現できる）

☐ This solid lines express the transmission lines.
(この実線は伝送線路を表している)

ポイント ☞n:expression 表現
☞目的語に wh-節を伴うことができる．形容詞としては「明白な急行の」．express train(急行列車)，express highway(高速道路)

extend
(extended, extended, extending)
〈他〉（範囲・方法・意味などを）拡張する，拡大する

☐ As a result, repeater spacing can be extended using conventional single-mode fibers.
(その結果, 中継器の間隔は, 従来の単一モードファイバを用いて拡大できる)

☐ It is worth mentioning that the technique can be extended to measure the relative permeability by modifying (12)-(14).
(その手法は(12)～(14)を変形することで，比透磁率を測定するために拡張できるということは言及しておく価値がある)

☐ The main results in this paper can be extended for modeling time series observed over a finite time interval,but we will not treat this issue here to simplify the presentation.
(本論文の主な結果は, 有限な時間間隔にわたって観測された時系列をモデル化するために拡張できるが，発表を簡単にするために，この問題はここでは取り扱わない)

☐ In Section 3 we will extend this method to three-dimensional translation.
(第 3 節では，この方法を三次元並進に拡張する)

☐ We extend the application of this active loop to a high temperature dielectric measurement.
(この能動ループの応用を，高温誘電体の測定に拡張する)

☐ In order to extend this combinational circuit methodology to sequential circuits, we use a model similar to that used for synchronous circuits.
(この組合わせ回路方法論を順序回路に拡張するために,同期回路用に使われるのに類似したモデルを使用する)

ポイント☞ [類]expand　n:extension 拡張　adj:extensive 広範囲にわたる

extract
(extracted, extracted, extracting)
〈他〉（信号・情報・データなどを）抽出する
extract A from B：BからAを抽出する

☐ The outline of an image is extracted by applying a differentiating mask on the image.
(画像の輪郭は，画像上に微分マスクを適用することで抽出される)

☐ The numerical charge control becomes applicable, and the parameters used in the analysis can be reliably extracted.
(その数値電荷制御は適用できるようになるため,解析で用いられるパラメータは確実に抽出できる)

☐ The clock, which is extracted from the data, is associated with high jitter.
(データから抽出されるクロックは高ジッタに関連している)

☐ The technique of synchronous demodulation extracts the required signal frequencies from the interference,giving a greatly improved signal-to-noise ratio at the transducer output.
(その同期検波法は，必要とする信号周波数を混信から抽出し,変換器の出力におけるS/N比を大幅に向上させる)

☐ An inherent problem of this approach is that it is impossible to extract the information as to how the layout was designed.
(このアプローチにつきまとう問題は,レイアウトはどのようにして設計されたかということについての情報を抽出することは不可能であるということである)

ポイント☞［類］sample　　n:extraction 抽出(物)

☞名詞としては「抽出物」.

F

fabricate

(fabricated, fabricated, fabribating)

〈他〉（装置・機械・製品などを）組み立てる，製作する

□ This modulator was fabricated in a 0.8 μ m CMOS process.

（この変調器は0.8 μ m CMOSプロセスで組み立てられた）

□ The 18-Mb synchronous DRAM has been fabricated with 0.5-μ m CMOS process technology.

（その18-Mb同期ＤＲＡＭは0.5-μ m CMOSプロセス技術で製作された）

□ The probes were fabricated from commercially available semirigid coaxial lines with outer diameters of 2.2 and 3.6 mm.

（そのプローブは，外径が2.2 mmと3.6 mmの市販の半硬質同軸線路を用いて組み立てられた）

□ The membrane for the thermal isolation of the heater was fabricated by selective etching.

（ヒータの断熱用の膜は，選択性エッチングで作られた）

□ Thin-film heads were fabricated using the multi-layered Co-Ni-Fe-Pd/Al_2O_3 films as magnetic cores.

（薄膜ヘッドは，磁心としてCo-Ni-Fe-Pd/Al_2O_3多層膜を用いることで製作された）

☐ We have fabricated and tested a number of integrated optical devices incorporating the features mentioned above.
（上記の特徴を含む，数多くの光集積デバイスを製作し，試験した）

☐ Fabry-Perot lasers fabricated from this wafer had an operating wavelength of 8,570 Å.
（このウェハから作られたファブリーペロレーザは，8,570 Åの動作波長を有する）

☐ In the case of the thin-film dc SQUID,one can make an integrated magnetometer by fabricating an Nb loop acrosse the spiral input coil.
（薄膜直流SQUIDの場合，らせん入力コイルの両端にNbループを組み立てることで，集積磁力計を作ることができる）

ポイント☞［類］build, make, manufacture, assemble, form, produce, prepare, construct　n:fabrication 組み立て・製作
☞「規格部品・材料などを用いて組み立てる」のニュアンスがある.

facilitate

(facilitated, facilitated, facilitating)

〈他〉（事を）容易にする，促進する

☐ This approach facilitated the implementation of VQ processors with a wide range of vector dimensions.
（このアプローチにより，高範囲のベクトル次元をもつVQプロセッサの実装が容易となる）

☐ This does not facilitate the design of a probe since each of these parameters has to be taken into account.
（これらのパラメータのどれもが考慮に入れなければならないので，これによりプローブの設計は容易とならない）

☐ This choice was made to facilitate comparison with the experimental results.
（実験結果との比較が容易となるように，これを選択した）

ポイント☞ ［類］ promote, hasten, further, accelerate, speed up　　n:facilitation 容易にすること・促進

fail

(failed, failed, failing)

〈自〉（機器などが）故障する，働かなくなる
（計画などが）失敗する
〈他〉 fail to do　～できない，～しそこなう

☐ This device fails at high humidities.
（この装置は高湿度で故障する）

☐ The lifetime of a transistor cannot be measured until it fails.
（トランジスタの寿命はそれが故障するまで測定できない）

☐ Everyone knows that sooner or later everything will fail.
（いつかはどのようなものでも故障するということは，誰もが知っていることである）

☐ Students failed mostly because they used a wrong method.
（学生は間違った方法を使用したため，ほとんど失敗した）

☐ It is demonstrated that the method fails to converge for the case of a discrete interface trap.
（その方法は離散界面トラップの場合には収束しないということが実証される）

☐ These methods work well for detecting disturbances,but can fail to capture fast changes.
（これらの方法は外乱を検出するためには十分な働きをするが，速い変化をとらえることはできない）

ポイント☞ ［反］ succeed(成功する)　［類］break down　　n:failure 故障・失敗

☞名詞としては「失敗」. without fail(必ず，まちがいなく)

feed

(fed, fed, feeding)

〈他〉 (信号・電力などを)送る，供給する，給電する
feed A into (to) B：AをBに送る，供給する
feed A with B：AにBを供給する
feed back：フィードバックする

☐ The comparator output signal is then fed to the switching section shown in Fig.8.
(コンパレータの出力信号は次に図8に示された切換え部に送られる)

☐ The frequency error is applied to a D/A converter whose output is fed to the integrator.
(周波数の誤差は，その出力が積分器に送られるD／A変換器に加えられる)

☐ The patch is assumed to be directly fed by means of a coaxial feeding point.
(パッチは同軸給電点によって直接給電されるものと仮定する)

☐ Memory elements sample a portion of an output signal and feed that information back into some of the input nodes.
(記憶素子は出力信号の一部をサンプリングし，その情報を入力ノードの一部にフィードバックする)

☐ A hemispherical dielectric resonator (DR) antenna fed by a microstrip line through a coupling slot on the ground plane has recently been studied.
(設置平面上の結合スロットを通したマイクロストリップ線路によって給電された半球型誘電共振器（DR）アンテナが，最近研究された)

☐ The measurement is usually done by means of a frequency counter fed with the filtered oscillator signal.
(濾波された発振器の信号を供給された周波数計数器によって，普通は測定が行われる)

☐ The position control is obtained by <u>feeding</u> back the position Θ, the angular velocity ω and the armature's current I.
(その位置制御は位置Θ，角速度ω，電機子電流Iをフィードバックすることで得られる)

☐ A speech signal is synthesized by <u>feeding</u> an excitation signal to an LPC synthesis filter.
(音声信号は，励起信号をLPC合成フィルタに送ることで合成される)

ポイント☞［類］provide, furnish, supply, serve
☞feeder は給電線・フィーダ，feedback はフィードバック・帰還．
feedback control(フィードバック制御)，negative feedback(負帰還)

fill

(filled, filled, filling)

〈他〉（物を）満たす，詰め込む
fill A with B：AをBで満たす，いっぱいにする

☐ The nonguiding region is <u>filled</u> with a dielectric medium of relative permittivity ε_2.
(非導波領域には，比誘電率がε_2の誘電媒質が充塡される)

☐ The tank was <u>filled</u> with distilled water at room temperature.
(タンクは室温において蒸留水で満たされた)

☐ Identical epoxy <u>filled</u> the gaps between the wafers and between the glass plates.
(全く同じエポキシが，ウェハ間，およびガラス板間の隙間を満たした)

☐ Analysis of electromagnetic wave propagation through a rectangular waveguide <u>filled</u> with a simple medium leads to the well-known dispersion relation.
(単純な媒質が充塡された方形導波管を通る電磁波伝搬の解析により，有名な分散関係が導かれる)

ポイント☞［反］empty(からにする)

☞名詞としては「容器一杯の量・充填物」.

filter

(filtered, filtered, filtering)

〈他〉（信号などを）ろ波する

□ The speech signal picked up by the microphone was filtered to avoid aliasing.
（マイクロホンで拾われた音声信号は，エイリアシングを避けるためにろ波された）

□ The output of the system was low-pass filtered with a four-stage Chebyshev filter at 15 MHz.
（システムの出力は，15 MHzで4段チェビシェフフィルタによって低域ろ波された）

□ The output from the gyro is filtered to reduce the $2\omega_m$ component and is then fed into a lock-in amplifier.
（ジャイロからの出力は，$2\omega_m$成分を少なくするためにろ波し，つぎにロックイン増幅器に送られる）

ポイント☞名詞としては「フィルタ・ろ波器」. filtering はフィルタリング.
digital filter(ディジタルフィルタ), high-pass filter(高域フィルタ)、low-pass filter(低域フィルタ), acoustic filter(音響フィルタ), optical filter(光学フィルタ)

find

(found, found, finding)

〈他〉（物・事を）見つける，見いだす，発見する，求める
find A (to be) B：AがBであることがわかる

□ Two types of chaotic attractors were found in this range of H.
（二つのタイプのカオス的アトラクタが，Hのこの領域で見いだされた）

□ The effective focal length and radius were found to be 7.75 cm and 0.60 cm,

respectively.
(実効焦点距離と半径は，それぞれが7.75 cmと0.60 cmであることがわかった)

□ It is found that the characteristic impedance increases as R increases.
(特性インピーダンスはRが大きくなるにつれて増大するということがわかった)

□ In the case of $y = A\sin\beta(x - ct)$,one can find a number of parameters that describe the motion.
($y = A\sin\beta(x - ct)$の場合，この運動を記述する多くのパラメータを求めることができる)

□ Frequency stabilized laser diodes have found many applications in lightwave communication and fiber sensor systems.
(周波数安定化レーザダイオードは，光波通信やファイバセンサシステムに多くの用途がある)

□ Piezoresistive strain gauges find widespread application in stress measurement in mechanical structures.
(圧抵抗歪みゲージは，機械構造物での応力測定に広く使われる)

□ The Newton iteration technique is used to find the solution.
(ニュートン反復法は，この解を求めるのに使われる)

□ The Fredholm integral equation of the second kind must be solved in order to find the potential distributions V_k on the surfaces S_k.
(表面S_k上の電位分布V_kを求めるために，第二種のフレドホルム型積分方程式を解かなければならない)

□ Two techniques have been used for finding the electric and magnetic fields from a known distribution of currents.
(既知の電流分布から電場と磁場を求めるのに，二つの手法が使われた)

ポイント☞ ［類］discover

☞「偶然あるいは研究などの結果，発見する」のニュアンスがある．名詞としては「発見」．finding も発見．

flow

(flowed, flowed, flowing)

〈自〉（電流・液体などが）流れる

☐ While the current i_Q is negative, it flows through the diode Di.
（電流i_Qがマイナスの間は，ダイオードDiを流れる）

☐ In a conventional generator,a plasma flows between two parallel rectangular electrodes.
（従来の発生器の場合，プラズマは二つの並列矩形電極間を流れる）

☐ The method is based on detecting the phase difference between the sinusoidal currents flowing through the metal electrodes.
（その方法は，金属電極を流れる正弦電流間の位相差の検出に基づいている）

☐ The variables x_{ij} is represented by the drain currents flowing MOS transistors.
（変数x_{ij}はMOSトランジスタを流れるドレイン電流で表される）

ポイント☞名詞としては「流れ」．flowchart(流れ図，フローチャート)，data flow(データフロー)，signal flow diagram(信号フローグラフ)

focus

(focused, focused, focusing)

〈他〉（光などが）集束する
（注意・関心などを）集中させる
〈自〉（注意・関心などが）集中する

☐ The polarized light is focused to the cleaved end face of the sample by a microscope lens.
（顕微鏡のレンズによって標本の劈開端面に偏光は集束する）

☐ Recently, the attention has been focused on the development of adaptive schemes

for plants with unknown relative degree.
（最近，相対次数が未知のプラントに対する適応方式の開発に関心が集まった）

□ So far, most research on learning control has been focused on manipulators that can move freely in the workspace.
（これまでは，学習制御の研究は,作業空間で自由に動けるマニピュレータに関心が集まってきた）

□ This problem can be overcome by using an offset reflector which focuses the reflected wave to the focal point outside the propagating zone, as shown in Fig.2.
（この問題は，図２に示すように,伝搬帯の外側の焦点に反射波を集束するオフセット反射器を用いることにより解決できる）

□ In this paper we focus on the shape optimization of layout elements.
（本論文では，レイアウト構成要素の形状最適化に焦点を当てる）

ポイント☞［類］concentrate
☞名詞としては「焦点(距離)」. focusing は焦点調節・ピント合わせ.

form
(formed, formed, forming)
〈他〉（物を）作る，形成する

□ These microcracks are formed due to the anisotropy of thermal expansion coefficients.
（これらの微小割れは，熱膨張係数の異方性のせいで形成される）

□ Thus a 10-15 μm thick layer of NbO_2 was formed on the surface of the NbO chip.
（したがって，10～15 μm厚のNbO_2層はNbOチップ表面上に形成された）

□ The gratings of Bragg reflectors are usually formed by holography.

(ブラッグ反射器の格子は，通常ホログラフィーによって形成される)

□ The coaxial conductors can be formed with a single-crystal HTS material.
(その同軸導体は単結晶HTS材料を用いて作ることができる)

□ The reflected waves form a conical wavefront.
(その反射波は円錐波面を作る)

□ In many electrical devices, for example transformers, the core will form a closed magnetic circuit.
(例えば変圧器といった多くの電気装置では，コアは閉磁気回路を形成する)

ポイント☞［類］build, make, manufacture, fabricate, assemble, produce, prepare, construct　n:formation 形成・構成，formula 公式　adj:formal 形式的な
☞「あるはっきりした形・構造に作り上げる」のニュアンスがある．名詞としては「形・型」.

formulate
(formulated, formulated, formulating)
〈他〉（事を）定式化する，公式化する，式で示す

□ An exact solution is formulated in the frequency domain using a spacial transform technique.
(空間変換法を用いて，周波数領域で厳密解を定式化する)

□ The design of these filters can be formulated as an optimization problem in a discrete parameter space.
(これらのフィルタの設計は，離散母数空間における最適化問題として定式化できる)

□ We formulate and solve an important problem known as the consistency between the data and a priori information.
(データ・事前情報間の一貫性として知られる重要な問題を定式化し，解く)

☐ Neglecting any magnetic saturation of the material,one can formulate an expression for the self inductance as a function of position.
(材料の磁気飽和を無視すれば,位置の関数として自己インダクタンスに対する式を表すことができる)

ポイント☞n:formulation 定式化・公式化

G

generalize

(generalized, generalized, generalizing)

〈他〉（事実・知識などを)一般化する

☐ This method can be generalized for the case of an inhomogeneous Green's function for nonlinear media.
(この方法は,非線形媒体のための非同次グリーン関数の場合に対して一般化できる)

☐ This procedure is generalized so that it can generate layouts under different parameter values and different design rules.
(さまざまなパラメータ値,およびさまざまな設計ルールのもとでレイアウトを生成するために，この手順を一般化する)

☐ We generalize the results in [2] to handle the m-input single-output LM model.
(m入力,単一出力のＬＭモデルを扱うために,[2]における結果を一般化する)

☐ The generalized transmission line model has been recently applied to the study of microstrip antennas.
(この一般化伝送線路モデルは,最近マイクロストリップアンテナの研究に適用された)

☐ We obtain a module generator by generalizing this procedure.
（この手順を一般化することで，モジュール発生器が得られる）

ポイント☞［反］specialize(特殊化する)　n:generalization 一般化

generate
(generated, generated, generating)
〈他〉（電気・熱・信号などを）発生させる，作り出す，生み出す

☐ The magnetic field was generated by a parmanent magnet.
（磁場は永久磁石によって発生した）

☐ The interference can be generated by electrostatic discharges, switches, or power lines.
（妨害は静電放電，スイッチ，あるいは電力線によって発生する可能性がある）

☐ High-frequency device noise is mainly generated by the random motion of electrons, and its power spectrum density depends on the local electron temperature.
（高周波装置の雑音は，主に電子の不規則運動によって発生するが，その雑音のパワースペクトル密度は局所的な電子温度に依存する）

☐ Harmonic incident longitudinal waves are generated with a contact piezo-electric transducer driven by a function generator.
（調和入射縦波は，関数発生器によって駆動される接触圧電変換器によって発生する）

☐ These hot carriers generate additional electron-hole pairs and lose their excess energy.
（これらのホットキャリアは余分な電子－正孔対を生み出し，過剰なエネルギーを失う）

☐ This is due to the saturation phenomenon which generates harmonics.
（これは，高調波を発生させる飽和現象による）

☐ These coils are commonly used to <u>generate</u> small, but highly uniform magnetic fields in the space between the coils.
(これらのコイルは，コイル間の空間に小さいが,非常に均一な磁場を発生させるのに通常用いられる)

☐ The distribution of heat <u>generated</u> by the solenoid was determined by measuring the electric field distribution.
(ソレノイドによって発生した熱の分布は,磁場分布を測定することによって求めた)

ポイント☞［類］produce　　n:generatin 発生

give

(gave, given, giving)

〈他〉（物・事を)与える，示す，施す
give ＡＢ／Ｂ to Ａ：ＡにＢを与える
give rise to：～を引き起こす，生じる

☐ The solution to this problem is <u>given</u> by the singular value decomposition.
(その問題の解は，特異値分解によって与えられる)

☐ The solution of Laplace's equation in the external region is <u>given</u> in the form of a trigonometric series.
(外部領域におけるラプラス方程式の解は，三角級数の形で与えられる)

☐ Prior to plasma etching,samples were <u>given</u> a standard organic solvent cleaning in an ultrasonic cleaner.
(プラズマエッチングの前に，超音波クリーナを使って,試料に標準有機溶剤洗浄を施した)

☐ This property <u>gives</u> the circuit a robustness agains bad data points.
(この性質があるため，回路は不良データ点に対してロバスト性を示す)

☐ We <u>give</u> a rigorous proof in the convergence of the learning process.

(学習過程の収束に対して厳密な証明を与える)

□ This means that it is difficult to give general guidelines for probe selection.
(このことは,プローブを選択するための一般的なガイドラインを与えることは困難であるということを意味している)

□ In solar cells,the nonuniform generation of carriers within the junction plane gives rise to a concentration gradient along the junction plane.
(太陽電池では,接合面内のキャリヤの不均一な発生は,接合面に沿って濃度勾配を引き起こす)

□ Thermal cycling under these conditions may give rise to constant strain amplitude.
(これらの条件の下での熱サイクリングは,一定の歪み振幅を引き起こすことがある)

□ In feedback amplifiers,knowledge of the poles and zeros of the forward amplifier gives insight into the performance and stability of the closed-loop feedback amplifier.
(帰還増幅器の場合,フォワード増幅器の極と零点についての知識は,閉ループ帰還増幅器の性能と安定性への見通しを与える)

ポイント☞［類］provide, apply
☞“give A B／B to A"において,前者ではBを,後者ではAが強調されている.

grow
(grew, grown, growing)

〈他〉(薄膜・結晶などを)成長させる
〈自〉(物・事が)発展する,増大する,成長する

□ CdS layers were grown at 300℃ on (100),(110),(111)A and (111)B substrates.
(300℃で,(100),(110),(111)A,(111)B基板上にCdS層を成長させた)

☐ This material was grown by molecular beam epitaxy on an Si-doped n-type GaAs substrage.
（分子線エピタキシーによって，Siドープn型GaAs基板上にこの材料を成長させた）

☐ The GaAs epitaxial layers on Si were grown using a standard MBE growth method.
（Si上のGaAsエピタキシャル層は，標準ＭＢＥ成長法を用いて成長させた）

☐ The use of capacitive proximity sensors has grown dramatically over the past five years.
（容量型近接センサの使用は，過去５年間にわたって劇的に増大してきた）

☐ Multimedia technology and the market for its use is growing and will continue to grow at a rapid rate.
（マルチメディア技術とその使用の需要は増大しており，急速に成長し続けるだろう）

☐ If a stress is applied to a sample containing a crack in an atmosphere containing water or water vapor, the crack will grow.
（水，あるいは水蒸気を含む雰囲気中で，亀裂のある試料に応力を加えるならば，その亀裂は成長するだろう）

☐ CdSe layers grown at 350℃ showed similar surface morphology to CdS.
（350℃で成長したCdSe層は，CdSと類似した表面形態を示した）

ポイント☞n:growth 成長
☞crystal growth(結晶成長)，epitaxial growth(エピタキシャル成長)

guarantee
(guaranteed, guaranteed, guaranteeing)
〈他〉（物・事を）保証する

☐ By using the above boundary layer control law, we can guarantee the attrac-

tiveness of the boundary layer.
(前述の境界層制御則を用いることで，この境界層が魅力的であることを保証できる)

□ The axial symmetry of the problem guarantees that the radial component of the magnetic current vanishes.
(その問題が軸対称であることにより，磁流の半径方向成分は消えるということが保証される)

□ The objective is to design a robust control which guarantees global stability and good tracking performance but requires only feedback of link position and velocity.
(この目的は，大域的安定性と優れたトラッキング性能を保証するが，リンクの位置と速度のフィードバックだけを必要とするロバスト制御器を設計することである)

□ There have been several ways to guarantee differentiability of robust control.
(ロバスト制御の微分可能性を保証する方法がいくつかある)

□ A robust control guaranteeing global stability was presented in [3].
(大域的安定性を保証するロバスト制御は[3]で示された)

ポイント ☞［類］assure, ensure, insure, warrant
☞「製品の品質や契約の履行に対して公式に責任をもつ」のニュアンスがある．名詞としては「保証」．

H

handle
(handled, handled, handling)
〈他〉（事を）処理する

☐ The case of P-MOS transistor can be handled in a similar way.
(P-MOSトランジスタの場合は，同様な方法で処理できる)

☐ The line connecting E_x and E_y nodes is handled as a line that propagation delay time is Δt.
(E_xとE_y結節点を連結している線は，伝搬遅延時間がΔtの線として処理される)

☐ The package can handle 100 W/cm^2 average power density.
(そのパッケージは100 W/cm^{2}の平均出力密度に対処できる)

☐ The analog Butterworth filter implementation with a 1 Hz cutoff frequency is a very economic and efficient way to eliminate high-frequency noise, but it can not handle the low-frequency noise in the 0 to 1 Hz range.
(遮断周波数が1 Hzのアナログ・バターワースフィルタの実施は，高周波雑音を除去するための非常に経済的で効率的な方法だが，0～1 Hz範囲の低周波雑音を処理することはできない)

ポイント☞［類］deal
☞名詞としては「取っ手・ハンドル」. 自動車のハンドルは (steering) wheel，自転車のハンドルは handlebar.

have

(had, had, having)

〈他〉（物を）もっている，所有する
（性質・特性などを）有する
have to do：～しなければならない

☐ Most rotating machines have a gap between the stator and the rotor.
(たいていの回転機には，固定子と回転子の間に間隙がある)

☐ The antenna has several advantages in mobile communication applications.
(そのアンテナは移動通信の用途でいくつかの利点をもっている)

☐ Symbolic analysis of electrical networks has several distinctive advantages compared with numerical analysis.
(電気回路網の記号解析には，数値解析に比較して，いくつかの独特の長所がある)

☐ It is shown that the technique has advantages over the well known technique of spatial filtering.
(その手法は，よく知られている空間フィルタリング法よりも有利であるということを示す)

☐ Laser techniques for generating and receiving ultrasound have potential for a wide range of non-contact ultrasonic measurements.
(超音波を発生させ，そして受信するためのレーザ手法には，広範囲の非接触超音波測定の可能性をもっている)

☐ In a modest videotelephony application, a video codec have to perform an 8×8 DCT at 10 frames per second.
(手頃なテレビジョン電話用途の場合，ビデオコーディックは10フレーム／秒で8×8 DCTを実行しなければならない)

☐ The He-Ne laser having 35 mW output power is used.
(35 mWの出力パワーを有するHe-Neレーザが用いられる)

ポイント☞［類］possess, own
☞進行形・受動態は不可．have only to(～しさえすればよい)，have nothing to do with(～と関係がない)，have an interest in(～に興味をもっている)，have difficulty in(～が困難である)

heat
(heated, heated, heating)
〈他〉（物を）加熱する，熱する

☐ The solder is heated to increase its fluidity while keeping the pressure P_l constant.

(圧力P_lを一定にしている間に流動度を増加するために,ハンダを加熱する)

□ These results suggest that the simi-cylindrical applicator can heat biological materials extremely efficiently.
(この半円筒状アプリケータは生体試料をきわめて効率よく加熱できるということを，これらの結果は示唆している)

□ The results of this investigation indicate that the solenoid can efficiently heat materials having conductivities comparable to those of biological materials.
(そのソレノイドは,生体試料の導電率に匹敵する導電率を有する材料を効率的に加熱できるということを，本研究の結果は示している)

ポイント☞［反］cool(冷却する)　adj:hot 熱い
☞名詞としては「熱」．加熱は heating．heater は加熱器・ヒータ．induction heating(誘導加熱), heat exchanger(熱交換器), heat-resistant glass(耐熱ガラス)，gas heater(ガスヒーター)

help

(helped, helped, helping)

〈他〉（物・事を）助ける
help（to）do：～するのを助ける
help A（to）do：Aが～するのを助ける

□ This helps confine the plasma to the region between the two electrodes.
(これは，二つの電極間の領域にプラズマを閉じ込める助けとなる)

□ A pipelining scheme in the read mode helps to shorten the access time by dividing the read-out operation into three stages.
(読み取りモードにおけるパイプライン方式は,読み出し動作を3ステージに分割することでアクセス時間を短縮するのを助ける)

□ These intermediate measurements helped us determine how advantageous it is to apply this method of separation to complex structures.
(これらの中間測定値は,この分離法を複合構造物に適用するのにどれくらい

有利かということを決定する助けとなる）

ポイント☞adj:helpful 役立つ
☞名詞としては「助け・助力」. cannot help～ing(～せざるを得ない)

hold
(held,held, holding)

〈他〉（物を）保持する
（物・事をある状態のまま）保つ
（データ・情報などを）保存（記憶，格納）する
hold together A／hold A together：Aをまとめる
〈自〉（式などが）成り立つ

□ The substrate was held on a stepper motor-driven X-Y table and the laser beam was focused onto the substrate surface.
(基板はステッパモータ駆動X-Yテーブル上に保持され, 基板の表面にレーザが当てられた)

□ The coolant temperature was held at −5℃.
(冷却剤の温度は−5℃に保たれた)

□ As the raindrop gets larger, the surface tension can no longer hold the drop together.
(雨滴が大きくなるにつれて, 表面張力はもはや雨滴をまとめることはできなくなる)

□ A single CD-ROM can hold the equivalent of approximately 270,000 pages of textual data or over 10,000 digitized images.
(１枚のCD-ROMは, およそ270,000ページのテキストデータ, あるいは10,000枚のディジタル化画像に相当する物を記憶できる)

□ It is easy to verify that (3.11) holds for $t+1$.
((3.11)は$t+1$に対して成り立つということを証明することは容易である)

ポイント☞［類］keep, maintain, preserve
☞名詞としては「保持」. hold time（保持時間）, sample-and-hold circuit（サンプルアンドホールド回路）

I

identify
(identified, identified, identifying)

〈他〉（物・事を）識別する，確認する，同定する

☐ These ghosts can be identified easily since they are independent of sample position.
（これらのゴーストはサンプルの位置とは無関係なので，容易に識別できる）

☐ Thermal runaway has been identified as one of the principal failure mechanisms in semiconductor devices.
（熱暴走は半導体素子の主な故障メカニズムの一つとして確認されている）

☐ Feature detection identifies special types of local patterns in the image.
（特徴検出は画像の特別な型の局所パターンを識別する）

☐ These calibration tests were essential to identify the test limits.
（これらの校正試験は，試験の限界をつまびらかにするのに不可欠だった）

☐ These measurements were used to identify the various types of noise in order to gain a better comprehension of the causes of oscillator instability.
（これらの測定値は，発振器の不安定性の原因をよりよく理解するために，さまざまな種類の雑音を識別するのに用いられた）

☐ The number of parameters to be identified is reduced by using the knowledge of

plant model structure.
(同定されるパラメータの数は, プラントモデルの構造の知識を用いることで減らす)

□ The method for identifying the location(s) of faulted subsystem(s) in the parallel communication system may be extended to detecting system faults for more complicated MIMO parallel nonlinear systems.
(並列通信システムにおける故障したサブシステムの位置を識別するための方法は, より複雑なＭＩＭＯ並列非線形システムのためのシステム故障を検出することに拡張できる)

ポイント☞［類］distinguish, differentiate, discriminate　　n:identification 識別・同定, identity 同一であること・一致
☞「間違いなくその物であると識別し、確認する」のニュアンスがある. system identification(システム同定), identification code(識別コード)

> **ignore**
> (ignored, ignored, ignoring)
>
> **〈他〉（物・事を）無視する**

□ All thin interface layers between the GaAs and the surface metal are ignored.
(GaASと表面金属の間の薄い界面層はすべて無視される)

□ For simplicity, we ignore the base current, and hence the collector current is equal to the emitter current.
(簡略化のために, ベース電流を無視するため, コレクタ電流はエミッタ電流に等しい)

□ This value may be ignored when calculating the equation of motion.
(この値は, 運動方程式を計算する時には無視してよい)

ポイント☞［類］neglect, disregard　　n:ignorance 無知　　adj:ignorant 無知の
☞「認めたくないことから目をそらして無視する」のニュアンスが

ある.

illustrate
(illustrated, illustrated, illustrating)
〈他〉(物・事を)示す，例示する，図示する

□ The results are illustrated in Fig.8.
(結果は図８に示す)

□ The equivalent bandwidth of the Butterworth low-pass filter is illustrated in Fig.2.
(バターワース低域フィルタの等価帯域幅を図２に示す)

□ In Fig.7 we illustrate a typical response resulting from one of these devices.
(図７に，これらの装置の一つから得られる代表的な応答を図示する)

□ Fig.1 shows the cross section of a typical three-phase VRM, which illustrates the essential features of the motors developed for these applications.
(図１に代表的な三相ＶＲＭの断面図を示すが，この図はこれらの用途のために開発された電動機のきわめて重要な特徴を示している)

□ The main purpose of this paper is to illustrate the effect of distortion on the phase error.
(本論文の主な目的は，位相誤差に及ぼす歪みの影響を示すことである)

ポイント☞［類］exhibit, show, display, reveal, demonstrate, indicate
n:illustration 実例・図解　adj:illustrative 実例となる
☞「図・実例などによって示す」のニュアンスがある。

implement
(implemented, implemented, implementing)
〈他〉(回路・システムなどを)実現する，実装する
(計画・プロジェクトなどを) 実行する，実施する

□ The delay line was implemented in a 0.8 μ m CMOS process.

（遅延線は0.8 μ m CMOSプロセスを用いて実現された）

□ The chips are implemented in 4 μ m NMOS technology.

（チップは4 μ m NMOS技術を用いて実装される）

□ VQ encoder algorithm for real-time image transmission and storage applications can be implemented using present day VLSI technology using high throughput systolic architectures.

（リアルタイム画像伝送・記憶用途のためのVQ符号器のアルゴリズムは，高スループットのシストリックアーキテクチャを用いた現在のVLSI技術を用いて実行できる）

□ The program implements a direct stability analysis algorithm.

（このプログラムは直接安定解析アルゴリズムを実行する）

□ This circuit does not correctly implement the state machine specified.

（この回路では特定の状態機械を正確に実現できない）

□ The method is very easy to implement as shown in Fig.1.

（その方法は図１に示すように，実行するのは非常に容易である）

□ A recently developed adaptive IIR filtering algorithm is used to implement an adaptive IIR echo canceller.

（最近開発された適応IIRフィルタリングアルゴリズムを，適応IIRエコーキャンセラを実現するのに用いる）

□ Fixed-point digital signal processors are suitable for implementing a large volume of products economically because they are usually much cheaper and faster than floating-point signal processors.

（固定小数点ディジタル信号プロセッサは，一般に浮動小数点信号プロセッサよりもずっと安価で高速なので，大量の製品を経済的に実現するのに適している）

ポイント☞［類］execute, carry out, run　　n:implementation 実現・実行

☞名詞(しばしば複数形)としては「道具・器具」.

improve

(improved, improved, improving)

〈他〉（物・事を）改善する，改良する

（性能・特性・価値などを）高める，向上させる

☐ The resolution of the FET spectrum was improved by decreasing the sampling rate.
（FETのスペクトルの分解能を，サンプリング周波数を低くすることで向上させた）

☐ The speed of other well-known Josephson circuits may not be improved by using high-temperature superconductors.
（他の有名なジョセフソン回路の速度は，高温超伝導体を用いても改善できないだろう）

☐ The position resolution can be further improved by increasing the modulation frequency.
（変調周波数を高くすることで，位置分解能はさらに改善できる）

☐ The linewidth of a conventional laser diode can be improved dramatically by coupling the laser to a diffraction grating.
（従来のレーザダイオードの線幅は，レーザを回折格子と結合することで劇的に向上できる）

☐ It is shown that a proper choice of the antenna shape can improve the cross-polarisation discrimination of the antenna system by more than 15 dB.
（アンテナの形状を適切に選択することで，アンテナシステムの干渉偏波の弁別を15 dB以上改善できる）

☐ A system of this type employs a semiconductor laser and a heterodyne interferometric setup to improve detection sensitivity.
（この種類のシステムは検出感度を高めるために，半導体レーザとヘテロダイン干渉装置を用いている）

□ There are two ways of improving the performance of neural networks as associative memory.
(連想メモリとしてのニューラルネットワークの性能を向上させる二つの方法がある)

ポイント☞［反］worsen(悪化させる)　［類］raise, enhance, upgrade
n:improvement 改善・改良・向上
☞「不備な点が改善されて前よりもよくなる」のニュアンスがある.

include
(included, included, including)

〈他〉（物・事を）含む，内蔵する

□ The phase angles of I_1 and I_2 should be included in these computations.
(これらの計算にI_1とI_2の位相角を含まなくてはならない)

□ When interface traps are included in the analysis, the Si-SiO_2 boundary condition becomes nonlinear.
(解析に界面トラップが含まれる場合，Si-SiO_2境界条件は非線形となる)

□ However, it should be noted that an attenuator can be included in an amplifier as a gain adjustment device.
(しかしながら，減衰器は利得調節装置として増幅器に内蔵できるという点に注目しなければならない)

□ The expression of the vector potential includes the modified Bessel function.
(そのベクトルポテンシャルの式は変形ベッセル関数を含んでいる)

ポイント☞［反］exclude（除く）　［類］contain, comprise, hold, involve
n:inclusion 含有　adj:inclusive 包括的な
☞「全体の一部分として含む」のニュアンスがある.

incorporate
(incorporated, incorporated, incorporating)

〈他〉 (物を)組み込む，含む
incorporate A in (into) B：AをBに組み入れる

☐ The four parameter listed above are <u>incorporated</u> in a CAD program as elements of the correlation matrix.
(上で列挙した4個のパラメータは，相関行列の要素としてCADプログラムに含まれる)

☐ If an electrooptic medium can be <u>incorporated</u>, then electrooptic tuning of the frequency of the laser will be possible.
(もし電気光学媒体を組み込むことができるならば，レーザの周波数の電気光学的同調が可能となるだろう)

☐ The power meter <u>incorporates</u> an InGaAs photodiode to detect optical wavelengths from 800 to 1600 nm.
(800～1600 nmの光波長を検出するために，この電力計はInGaAsフォトダイオードを内蔵している)

☐ This is achieved by <u>incorporating</u> each servomotor into its own servomechanism, as shown in Fig.1.
(このことは，図1に示すように，各サーボモータを各自のサーボ機構に組み入れることによって達成される)

ポイント☞n:incorporation 合併・会社
☞形容詞としては「会社組織の・合併した」.

increase
(increased, increased, increasing)
〈他〉 (大きさ・数量・程度などを)増やす，増加させる
(能力などを)向上させる
〈自〉 (大きさ・数量・程度が)増える，増大する

☐ The accuracy of position measurement can be significantly <u>increased</u> by a calibration of the whole system.
(位置測定の精度は，全システムを校正することで著しく向上できる)

□ The temperature can be further increased by increasing the current.
(温度は電流を増加することで，さらに上昇させることができる)

□ This signal averaging procedure significantly increases the signal-to-noise ratio.
(この信号平均化法は，信号対雑音比を著しく改善する)

□ However, the parasitic capacitance C_p associated with R_1 and R_2 also increases.
(しかしながら，R_1とR_2に関連する寄生容量C_pもまた増加する)

□ Device density has increased at a rate of about 40 percent per year.
(デバイスの密度は，年間約40%の割合で増大してきた)

□ This voltage will increase with increasing field strength, which will in turn decrease the capacitance.
(この電圧は場の強度が増大するにつれて増加し，その結果静電容量を減少させることになる)

□ The more the performances of microprocessor increased,the more the difficulties of peripheral circuit designing increased.
(マイクロプロセッサの性能が向上するにつれて，周辺回路の設計はますます困難となる)

□ Recently, the need for data communication with very low bit error rate through mobile radio channels is increasing.
(最近，移動無線チャネルを介したビット誤りが非常に小さいデータ通信の必要性が増大している)

ポイント☞ ［反］decrease（減少する）　［類］augment, enhance, rise
adj:increasing ますます増える　adv:increasingly ますます

indicate
(indicated, indicated, indicating)

〈他〉（物・事を）示す，表す

☐ The three main leakage current paths are indicated by the arrows in Fig.4.
（三つの主な漏れ電流の経路を，図4中の矢印によって表す）

☐ A comparison of Figure 8 to Figure 9 indicates that the two fibers generally have similar responses.
（図8と9との比較で，この2本のファイバは一般に似た応答を有しているということが明らかとなる）

☐ The frequency characteristics indicate that the proposed SR enhancement technique never degrades the frequency characteristic of the opamp.
（その周波数特性は，提案されたSR強調法は，演算増幅器の周波数特性を決して劣化させないということを示している）

☐ This section indicates how to prove Theorems 1 and 2.
（この節では，定理1と2をどのようにして証明するかということについて述べる）

ポイント☞［類］exhibit, show, display, reveal, demonstrate, illustrate
n:indication 指示・兆候　　adj:indicative 示している
☞「ある事柄の意味・内容などを明確に示す」のニュアンスがある．

induce
(induced, induced, inducing)
〈他〉（電流・磁気などを）誘導する

☐ The electric fields, which can be induced at air-substrate interfaces during reflection of the EMPs, will be evaluated.
（EMPの反射の間に，空気・基板界面で誘導される電場を評価する）

☐ This magnetic flux will induce an electric voltage v(*t*) proportional to the rate of change of the flux.
（この磁束は，磁束の変化率に比例する電圧v(*t*)を誘導する）

☐ The modeling of voltages induced on overhead wires by external electric and magnetic fields is an important area of electromagnetics research.
(外部電界および磁界によって架空電線に誘導される電圧のモデリングは，電磁気学研究の重要な領域の一つである)

ポイント ☞n:inducement 誘導, induction 誘導
☞electromagnetic induction(電磁誘導)

influence
(influenced, influenced, influencing)
〈他〉（物・事に）影響を及ぼす

☐ The signals are especially influenced by background illumination and the rough surface of the particles.
(その信号は，背景照度と粒子のざらざらした表面にとりわけ影響を受ける)

☐ τ is no longer influenced by electrons trapped in shallow electron traps.
(τ は浅い電子トラップに捕獲された電子には，もはや影響されない)

☐ We have recently found that such degradation is strongly influenced by the substrate temperature.
(そのような劣化は基板温度に強く影響されるということを最近見いだした)

☐ There are a number of factors that influence the behavior of the simulated electromagnetic field.
(そのシミュレートされた電磁場の挙動に影響を及ぼす多くの要因がある)

ポイント ☞［類］affect　adj:influential 影響力のある
☞名詞としては「影響」. under the influence(～の影響を受けて)

integrate
(integrated, integrated, integrating)
〈他〉（回路などを）集積（化）する
（物を）まとめる，一体化する，統合する

integrate A with B：AをBと統合する
integrate A into B：AをまとめてBにする

□ InGaAsP Bragg waveguides were monolithically integrated with 1.5 μm InGaAsP laser diodes.
（InGaAsPブラッグ導波管を1.5 μmInGaAsPレーザダイオードとモノリシック集積化した）

□ Fig.11 shows a microphotograph of the circuit, which is integrated in a standard BiFET process.
（図１１は回路のマイクロ写真を示しているが，この回路は標準BiFETプロセスにより集積化している）

□ Because the transducer is fabricated using silicon technology, peripheral circuits, such as drivers, delay lines and analog signal processors, can be integrated on the same die.
（その変換器はシリコン技術を用いて組み立てられているので，ドライバ，遅延線，アナログ信号プロセッサといった周辺回路は同じダイ上に集積化できる）

□ The instrumentation amplifier and the bandpass filter have been integrated for that purpose in a bipolar process.
（その目的のために，計測増幅器と帯域通過フィルタはバイポーラプロセスで集積化された）

□ When semiconductor devices are to be integrated into the microstrip structure, silicon is often used as the substrate material.
（半導体デバイスは集積化されてマイクロストリック構造とする場合，シリコンはしばしば基板材料として用いられる）

□ In its simplest form, multimedia integrates video, audio, text, image and graphics via CD-ROMs, video tape, or video servers.
（その最も簡単な形式の場合，マルチメディアはビデオ，オーディオ，テキスト，画像，グラフィックスをCD-ROM，ビデオテープ，あるいはビデオサー

バによって統合する）

☐ Sigma-delta circuits are very attractive analogue-to-digital converters because they achieve high accuracy without the need to integrate critical analogue components.
（シグマ－デルタ回路は非常に魅力的なアナログ－ディジタル変換器である．というのは，重要なアナログ部品の一体化を必要とせずに，高精度を達成するからである）

☐ Integrating A/D converters are widely used in digital multimeters and data acquisition systems.
（A/D変換器の集積化は，ディジタル・マルチメータとデータ収集システムで広く用いられる）

ポイント☞n:integration 集積化　　adj:integrated 集積された
☞large scale integrated circuit(大規模集積回路), monolithic integrated circuit(モノリシック集積回路)

interact
(interacted, interacted, interacting)
〈自〉（物と）相互作用する
interact with Ａ：Ａと相互作用する

☐ We will assume that the conductors do not interact thermally.
（この導体は熱によっては相互作用しないと仮定する）

☐ Magnetic fields interact with these materials.
（磁場はこれらの材料と相互作用する）

☐ This technology allows humans to interact with machines in a more natural manner by means of speech recognition and speech synthesis.
（この技術は，人間が音声認識と音声合成によってより自然な方法で機械と相互作用できるようにする）

☐ In this approach, the radiation fields interacting with the molecular medium are treated classically.
（この手法では，分子媒体と相互作用する放射場は古典的に扱われる）

ポイント☞n:interaction 相互作用　　adj:interactive 相互作用の
☞electromagnetic interaction(電磁相互作用)

interfere
(interfered, interfered, interfering)
〈自〉（物・事を）妨害する，干渉する
interfere with A：Aを妨げる

☐ Other parts of the human body were not taken into consideration because they do not interfere with the radiated fields.
（人体の他の部分は，放射場を妨げないので考慮しなかった）

☐ Clearly, a slow thermal drift of the sensor should not interfere with conductivity measurements.
（センサの遅い熱ドリフトが導電率の測定を妨げてはならないのは明らかである）

☐ The reference antenna is set so as not to interfere to electric field of emission sources.
（その基準アンテナは放射源の電界と干渉しないように取り付ける）

ポイント☞n:interference 妨害・干渉
☞electromagnetic interference(電磁障害・電磁干渉), interference fringe(干渉縞), interferometer は干渉計.

investigate
(investigated, investigated, investigating)
〈他〉（物・事を）調べる，研究する，調査する

☐ Fibre ring resonators have been extensively investigated by several researchers.

(ファイバリング共振器は,幾人かの研究者によって広範囲にわたって研究された)

☐ A number of researchers have investigated the reduced-order estimation problem.

(多くの研究者により，低次元推定問題は研究されてきた)

☐ The paper investigates in detail the posible application of neural networks to direct model reference adaptive control.

(本論文では,直接モデル規範適応制御へのニューラルネットワークの実現可能な応用について詳細に調べる)

☐ We investigate the properties of the closed loop using the proposed controller structure.

(提案された制御器構造を用いて，閉ループの特性を調べる)

☐ It is of interest to investigate the temperature dependence of the Faraday effect in low-birefringence fibers.

(低複屈折ファイバにおけるファラデー効果の温度依存性を調べることは興味深いことである)

☐ The technique for accomplishing this result is being investigated.

(この結果を達成する手法を研究中である)

☐ We are currently investigating the transient convergence properties of the multistage CMA adaptive beamformer.

(現在，多段ＣＭＡ適応ビームフォーマの過渡収束特性について研究している)

ポイント☞［類］study, examine, research　　n:investigation 調査・研究
adj:investigative 調査の
☞「事実関係などを詳細に徹底的に調べる」のニュアンスがある.

involve

(involved, involved, involving)

〈他〉 (物・事を)含む，伴う

□ The second categroy of speech technology applications in automobiles involves speech recognition systems.
(自動車での音声技術用途の第二のカテゴリーには，音声認識システムが含まれる)

□ The basic algorithm involves mainly addition, subtraction, comparison, and logic decision.
(その基本アルゴリズムには，主に加法，減法，比較，論理的決定が含まれている)

□ The Lyapunov analysis implies that the presently analyzed EEG data involves chaos which may be characterized by fractal dimensions of their attractors in phase space.
(今解析されたＥＥＧデータには，相空間中のアトラクタのフラクタル次元によって特徴づけできるカオスが含まれるということを，リャプノフ解析によって示す)

□ This process involves no matrix manipulation and uses only real arithmetic.
(この過程には行列処理が含まれておらず，実算術のみを用いている)

ポイント☞[類]include, contain, comprise, hold　n:involvement関与　adj:involved 複雑な・関係のある

isolate

(isolated, isolated, isolating)

〈他〉 (回路などを)絶縁する
(物を)分離する，隔離する
isolate A from B：AをBから分離する，隔離する

□ In order to keep the power consumption of the chip low, the heater is thermally

isolated from the heat sink by a membrane.
(チップの電力消費を小さくしておくために，膜によってヒータをヒートシンクから断熱する)

☐ However, circuits are rarely isolated from nearby objects that affect the radiated fields.
(しかしながら，回路は放射場に影響を及ぼす近くの物体からめったに隔離されない)

☐ The laser was optically isolated from the sample to eliminate any effect of the reflected light on its spectrum.
(反射光のスペクトルに及ぼす影響を除去するために，レーザはサンプルから光学的に分離された)

☐ The device was electrically isolated.
(デバイスを電気絶縁した)

ポイント☞n:isolation 絶縁・分離
☞isolator は絶縁体

K

keep

(kept, kept, keeping)

〈他〉 (ある動作・状態を)続ける，維持する，保つ
keep A B：AをBにしておく
keep A from doing：Aに～させないようにする
〈自〉 (物・事が)～し続ける
keep doing：～し続ける

□ The magnetizing current I_{MT} of the transformer is kept constant.
(変圧器の磁化電流I_{MT}は一定に保たれる)

□ The alloys have a high saturation polarization,and their magnetostriction can be kept very low with a material of appropriate composition.
(その合金には高飽和分極という特性があり，適切な組成の材料を用いてその合金の磁気歪みを非常に小さくしておくことができる)

□ Spill prevention devices are high-level alarms that help keep above-ground storage tanks from spilling over.
(流出防止装置は，地上貯蔵タンクがあふれ出るのを防ぐのに役立つ高水準警報装置である)

□ The suppliers keep providing a steady flow of off-the-shelf monolithic and hybrid devices with high power output.
(その部品供給元は，電力出力が大きい出荷待ちのモノリシックおよびハイブリッドデバイスの定常的な流れを提供し続けている)

□ To avoid coupling effects we had to keep chip area relatively large.
(結合の影響を避けるために，チップの面積は比較的大きくしておかなければならなかった)

□ The principle of the mass-flow sensor is based on the convective heat transfer from a heater kept at a constant temperature T_h.
(この質量流量センサの原理は，一定の温度T_hに保たれたヒータからの対流熱伝達に基づいている)

ポイント☞［類］maintain, hold, preserve, continue
☞"keep A B"において，Bは形容詞・現在分詞・過去分詞．

know

(knew, known, knowing)

〈他〉(物・事を)知っている，わかっている
know that節：～ということを知っている

know A to be (as) B：AがBだと知っている

□ The parameters Θ and *H* are <u>known</u> beforehand, while *K* and *M* are unknown and must be determined.
(パラメータΘと*H*はあらかじめわかっているが，その一方，*K*と*M*は未知なので，決定しなければならない)

□ It is well <u>known</u> that the performance of a reflector antenna is critically dependent on the reflector surface accuracy.
(反射型アンテナの性能が反射器表面の精度に大きく左右されるということは，よく知られている)

□ From Theorems 1 and 2 we <u>know</u> that the zero state response of $y(t)$ is expressible as the solution of the following first order differential equation.
(定理１と２より，$y(t)$のゼロ状態応答は，次の一次微分方程式の解で表すことができるということがわかる)

□ Non-Gaussian impulsive noise is <u>known</u> to be one of the major sources of errors in digital transmission systems.
(非ガウスインパルス雑音は,ディジタル伝送システムにおける誤りの主な源の一つであるということは知られている)

□ Numerical methods are generally <u>known</u> to be effective for the analysis of complicated systems.
(数値法は複雑なシステムの解析では効果的であるということは一般に知られている)

□ This is <u>known</u> as the locally optimal detector.
(これは局部最適検出器として知られている)

ポイント☞［類］comprehend, understand　　n:knowledge 知識
☞進行形は不可．that節以外に wh-節や if節も可能である．“know A to be B"において，Bは形容詞・名詞だが，不定詞を直接目的語にとることはできない．

☞knowledge base(知識ベース)，knowledge engineering(知識工学)

L

lead

(led, led, leading)

〈自〉（物・事が)至る，通じる

lead to A：(結果として)Aとなる，Aに至る，Aをもたらす

☐ The combination of the FEM and MoM techniques leads to an improvement in the numerical accuracy of the solution.

(FEM法とMoM法を組み合わせることで，この解法の数値精度が向上する)

☐ The remarkable advances in speech processing in the past two decades have led to the development of several pitch detection algorithms.

(過去20年間の音声処理の著しい進歩により，いくつかのピッチ検出アルゴリズムが開発された)

☐ Polarization drifts can lead to variations in the relative amplitudes of the scattered beams.

(偏光ドリフトは，散乱ビームの相対振幅の変動をもたらすことがある)

☐ Magnetization reversal, which leads to magnetic domain noise, does not occur.

(磁区雑音をもたらす磁化反転は起こらない)

☐ The dynamic programming equation leads to the following optimality conditions:

(動的計画法方程式は次の最適性条件をもたらす)

☐ In recent years, the increased number of nuclear power plants has led to the

inevitable problem of how to dispose of radioactive waste.
(近年，原子力発電所が増加したことで，放射性廃棄物をどのようにして処理するかという不可避の問題をもたらした)

ポイント☞［類］guide,result in　　adj:leading 主要な・第一級の
☞名詞としては「先導・指導」.

lie
(lay, lain, lying)
〈自〉（事実・理由・欠陥などが）ある
lie in A：Aにある

□ A foreseeable disadvantage lies in the long-term mechanical stability of the assembly.
(予見できる不利な点がアセンブリの長期機械的安定度にある)

□ The computational power of a neural networks lies in the high degree of local and global connectivity.
(ニューラルネットワークの計算能力は，高度な局所的および大域的連結性にある)

□ The utility of neural network lies in their ability to infer and generalize a mapping function from examples.
(ニューラルネットワークの有用性は，例から写像関数を推論し，一般化する能力にある)

□ The lifetimes obtained from (8) lie on curve LL´ in Fig.5.
((8)から得られた寿命は図5の曲線LL´上にある)

ポイント☞lay(横たえる)－laid－laid の変化と混同しないこと.

limit
(limited, limited, limiting)
〈他〉（物・事を）制限する

limit A to B：AをBに制限する

☐ The performance is mainly limited by the characteristics of the rubber.
(その性能は主としてゴムの特性によって制限される)

☐ The bandwidth of a receiver is no longer limited by the amplifier bandwidth.
(受信機の帯域幅は，もはや増幅器の帯域幅によって制限を受けない)

☐ The performance of the magnetometer is limited by noise in the sidebands which reduces the signal-to-noise ratio of the magnetic signal to be detected.
(磁力計の性能は,検出される磁気信号のS/N比を低減する側波帯中の雑音によって制限される)

☐ At high output currents the THD is limited to 0.9%.
(高出力電流では，THDは0.9%に制限される)

☐ This dependence affects the performance of A-to-D and D-to-A converters and limits their use in many applications.
(この依存性はA/DおよびD/A変換器の性能に影響を与え，多くの用途でのそれらの使用に制限を与える)

☐ This factor limits the modulation to a relatively low frequency.
(この係数により，変調は相対的に低い周波数に制限される)

☐ At present, the computational capability of control centres has limited security analysis to steady state calculations.
(現在，制御センターの計算能力のため,セキュリティ解析は定常状態における計算に制限されている)

ポイント ☞［類］restrict, confine　n:limitation 制限　adj:limited 限られた，limiting 制限する

☞「限界を決めて，それを越えないようにする」のニュアンスがある．名詞としては「限界・極限」．upper limit(上限), lower limit(下限)，limit value(極限値)

list

(listed, listed, listing)

〈他〉（物・事を）一覧表にする，列挙する

□ The values of these parameters are listed in Table 1.
（これらのパラメータの値を表１に一覧する）

□ The parameters of the thin-film medium and the heads used for magnetic recording are listed in Table Ⅰ and Ⅱ, respectively.
（薄膜媒体と磁気記録用に使われるヘッドのパラメータを，それぞれ表Ⅰと表Ⅱに一覧する）

□ Table 2 lists the thermal properties of silicon and several other materials used for carriers.
（表２に，キャリアに使われるシリコンと他のいくつかの材料の熱特性を列挙する）

□ In this section we list some frequently used notation and briefly review the statistical concepts used throughout the paper.
（この節では，しばしば使用される記号を列挙し，本論文を通して使用される統計的概念を簡単に概説する）

□ Each of the materials listed has been fabricated in thin-film form.
（列挙してあるどの材料も，薄膜の形で製造された）

ポイント☞名詞としては「表・一覧表・リスト」.

lower

(lowered, lowered, lowering)

〈他〉（速度・温度・程度などを）下げる，低下させる

□ As the thickness is decreased to less than 0.5 μm, H_{ch} is lowered abruptly.
（厚みが0.5 μm以下まで減少するにつれて，H_{ch}は急速に下がる）

□ As the probe was lowered, cooling curves of resistance versus temperature, ranging from 320 K to 4 K, could be obtained because of the existence of a temperature gradient within the Dewar vessel.
（プローブが下がるにつれて，320 K～4 Kの範囲にある抵抗対温度の冷却曲線を得ることができた。というのは，デュアー瓶内に温度勾配があるからである）

□ Although some films started to show a drop in resistance as the temperature was lowered, no films exhibited zero resistance.
（膜のなかには，温度が下がるにつれて抵抗降下を示し始めるものもあったが，零点抵抗を示す膜はなかった）

□ The two driver ICs can lower the power consumption of STN LCD modules.
（二つのドライバICにより，STN LCDモジュールの消費電力を低減できる）

□ These defects degrade circuit performance and lower the yield of ICs.
（これらの欠陥により回路の性能は劣化し，ICの歩留りは低くなる）

□ Lowering the transistors' V_{TH} helps to improve circuit speed with a low supply voltage.
（トランジスタのV_{TH}を低下させることは，供給電圧が低い回路の速度を向上させる助けとなる）

ポイント☞［反］heighten(高くする)，lift(上げる)

M

maintain

(maintained, maintained, maintaining)

他〉（状態・温度・圧力・性能などを）維持する，保持する

□ The SQUID was maintained at 4.2 K.
（SQUIDは4.2 Kに維持された）

□ The output voltage can be maintained for up to 1 second at 25℃.
（その出力電圧は25℃で最高1秒間維持できる）

□ The resonant frequency is maintained at a constant 3.7428 GHz, and the directivity increases as A increases until the side lobe is formed.
（その共振周波数は一定周波数3.7428 GHzに維持され，指向性はサイドローブが形成されるまで，Aの増加とともに向上する）

□ Run-length codes maintain a certain minimum and maximum time difference between transitions in the modulation signal $w(t)$.
（ランレングス符号は，変調信号$w(t)$の遷移する間のある最小，および最大時間差を保持する）

ポイント☞［類］keep, hold, preserve　　n:maintenance 維持・保守・メンテナンス

☞preventive maintenance(予防保全)，maintenance-free(保守不要の)

make

(made, made, making)

〈他〉（物を）作る，製作する
（物・事を）～にする
（物・事が）～させる
make A of B：AをBで作る
make A from B：AをBで作る
※ of はBが本質的に変化しない場合に，from はBが変化してもとの形をとどめていない場合に用いる
be made up of A：Aからなる，構成される
make A B：AをBにする
※ Bは名詞，形容詞，分詞
make A do：Aに～させる

make＋動詞より派生した名詞：～をする，行う
make use of A：Aを利用する
make up A：Aを構成する

☐ Various cobalt-containing alloys were made in the form of amorphous ribbons.
（種々のコバルト含有合金がアモルファスリボンの形で作られた）

☐ The cover is made of plastic.
（そのカバーはプラスチック製である）

☐ C_p is quite large because R_1 and R_2 are made of doped polysilicon.
（R_1とR_2はドープされたポリシリコンで作られているので，C_pはかなり大きい）

☐ These matching transformers can be made of dielectric materials or of ferroelectric materials.
（これらの整合変圧器は誘電材料，あるいは強誘電材料で作ることができる）

☐ Samples were 31 mm long with an apex angle of 35° and were made of magneto-strictive polycrystalline ferrite.
（試料は長さが31 mm，頂角が35度で，磁気歪み多結晶フェライトで作られている）

☐ The driving element of the piezo-motor is made from piezoelectric bilayer structure actuator.
（圧電モータの駆動部品は，圧電二層構造アクチュエータで作られている）

☐ Early high *Tc* SQUIDs made from YBCO had several problems.
（YBCOで作られた初期の高*Tc* SQUIDには，いくつかの問題点があった）

☐ Lead-Zirconate-Titanate (PZT) is made up of a $PbZrO_3/PbTiO_3$ mixture.
（ジルコン酸チタン酸塩 (PZT) は$PbZrO_3/PbTiO_3$混合物からなる）

☐ It is made up of the divider, the gate, the control circuits, the binary counters, and the latches.

(それは分周器，ゲート，制御回路，２進計数器，ラッチから構成される)

☐ The magnetic domains were made visible with a polarizing microscope by means of the Kerr effect.
(カー効果を用いた偏光顕微鏡によって磁区を視覚化した)

☐ These advantages make the WD switching system attractive for future network expansion.
(これらの利点があるため，将来のネットワークの拡張に対してＷＤ交換システムは魅力的となっている)

☐ In order to make a noise figure of IF circuit low,the IF amplifier must be low noise and high gain.
(IF回路の雑音指数を小さくするには，IF増幅器は低雑音で高利得でなければならない)

☐ This environment makes it possible to efficiently perform the hardware verification and the software verification at both board-level and system-level simulation.
(本環境では，ボードレベルないしシステムレベルでのシミュレーションでのハードウェア検証とソフトウェア検証を効率的に行うことができる)

☐ Although these global feedback paths are all negative,it is possible to make the SC circuit oscillate if the global feedback is large enough.
(これらの大域的フィードバック経路がすべて負ではあるが，大域的フィードバックが十分に大きければ，SC回路を発振させることは可能である)

☐ Attempts have been made to control the characteristics of these devices by supplying an electric current to the superconductive thin wires.
(電流を超伝導細線に供給することで，これらのデバイスの特性を制御するという試みが行われた)

☐ Discussion is made on the behavior of the oxide film over the surface of the aluminum alloy before and after baking.

（アルミニウム合金表面の酸化膜のベーキング前後の挙動について議論する）

☐ Noise measurements have been made in both the frequency and time domains.
（雑音の測定が，周波数領域と時間領域の両方で行われた）

☐ The decision to use a signal processor for the control was made at the beginning of this project in 1988.
（制御のために信号プロセッサを使用するという決定は，1988年にこのプロジェクトの初めに行われた）

☐ To decode the signals, HDTV receivers will make extensive use of digital image processing and frame memories.
（信号を復号するために，HDTV受信機はディジタル画像処理とフレームメモリを広範に利用する）

☐ The function of the superconducting FET is similar to that of a conventional semiconductor FET, but it makes use of a conduction mechanism peculiar to superconductivity.
（超伝導FETの機能は従来の半導体FETの機能に似ているが，超伝導に特有の伝導機構を利用している）

☐ One of the most important routines making up the control program is the FFT routine.
（制御プログラムを構成している最も重要なルーチンの一つは，FFTルーチンである）

ポイント☞ ［類］build, manufacture, fabricate, assemble, form, produce, prepare. construct

match

(matched, matched, matching)

〈他〉（物・事と）合う，一致する，適合する
（回路などを）整合する
match A to (with) B：AをBと整合させる

☐ This design is well matched at up to 15 dB attenuation.
(この設計は，最大15 dBの減衰まで十分適合する)

☐ Center bandwidth of the polarizer was chosen to match optical source center bandwidth.
(光源の中心帯域幅に一致させるために，偏光子の中心帯域幅を選んだ)

☐ A thin-film polysilicon low-temperature process has been developed which is well matched to the interated-scanner application.
(集積スキャナ用途に十分に適合する，薄膜ポリシリコン低温プロセスが開発された)

☐ This value can be matched theoretically using $\alpha_e = 2.87 \times 10^8 \mathrm{Vcm}^{-1}$,and $E_0 = 4 \times 10^6 \mathrm{Vcm}^{-1}$.
($\alpha_e = 2.87 \times 10^8 \mathrm{Vcm}^{-1}$と$E_0 = 4 \times 10^6 \mathrm{Vcm}^{-1}$を用いれば，この値は理論的に一致する)

☐ If Z_L is equal to the transmission line characteristic impedance Z_0,then Fig.2 represents a transmission line that is matched.
(もしZ_Lが伝送線路の特性インピーダンスZ_0と等しければ，図２は整合する伝送線路を表すことになる)

☐ The technique was used to develop a balun circuit to match a 900-MHz slot antenna to 50 Ω.
(その手法を900-MHzスロットアンテナを50 Ωに整合させるためのバラン回路を開発するために用いた)

☐ The choke helps to match the antenna to the coaxial transmission line.
(そのチョークはアンテナを同軸伝送線路に整合させる助けとなる)

ポイント☞［類］agree, accord, conform, coincide, concur, correspond, fit
n.(adj.):matching 一致・マッチング(している)
☞名詞としては「一致・適合」. pattern matching(パターン照合), template matching(テンプレートマッチング)

measure

(measured, measured, measuring)

〈他〉（長さ・量・大きさなどを）測定する

（〜の）寸法をしている

□ The amplitude of the harmonic at 30 MHz is measued with a spectrum analyzer.
（30 MHzでの高調波の振幅をスペクトルアナライザで測定する）

□ The transit time of the optical pulses at different wavelengths is measured over the wavelength region 1.1-1.7 μm.
（さまざまな波長の光パルスの走行時間を，1.1〜1.7 μmの波長領域にわたって測定する）

□ The transmission coefficient S_{21} was measured to determine the crosstalk voltage transfer functions.
（クロストーク電圧伝達関数を決定するために，透過係数S_{21}を測定した）

□ The maximum controllable current of the depletion-mode thyristor has been experimentally measured as a function of the device structrue parameters and the ambient temperature.
（デプレッション型サイリスタの制御可能な最大電流は，素子構造のパラメータと周囲温度の関数として実験的に測定された）

□ Detectors 1 and 2 measure the light intensities S_1 and S_2.
（検出器１と２は光度S_1とS_2を測定する）

□ The digital output circuit measures the signal amplitude in the resonator.
（ディジタル出力回路は，共振器の信号振幅を測定する）

□ The completed chip measures 7mm by 10mm in an 0.8 μm CMOS technology.
（その完成したチップの大きさは，0.8 μm CMOS技術を使って7×10mmである）

□ What is the best way to measure pressure ?

(圧力を測定する最もよい方法は何か)

□ Figure 3 shows pressure waveform measured by the miniature hydrophone.
(図3は小型ハイドロホンで測定された圧力波形を示している)

□ A simple method of measuring the propagation constant at a single frequency is described.
(単一周波数で伝搬定数を測定する簡単な方法について述べる)

ポイント☞n:measurement 測定(値)
☞名詞としては「測定・測度・寸法・手段」. meaure space(測度空間)

minimize

(minimized, minimized, minimizing)

〈他〉(数量などを)最小にする

□ Baseline drift is minimized by controlling the drift sources.
(基線のドリフトはドリフトの源を制御することで最小にする)

□ This Figure shows that estimate error can be minimized by judicious selection of plant.
(この図は，推定値の誤差が適切な判断でプラントを選択することで,最小にできるということを示している)

□ Back propagation is a gradient descent method which may minimize the difference between input pattern vectors and output pattern vectors.
(逆伝搬法は,入力パターンベクトルと出力パターンベクトルの間の差を最小にできる，一種の傾斜降下法である)

□ These methods provide long-term safety and minimize storage space.
(これらの方法は，長期安全性を実現し，記憶空間を最小にする)

□ In this steady state, we can assume $\nabla_v \varepsilon = \nabla_w \varepsilon = 0$ to minimize the cost function.

（この定常状態の場合，費用関数を最小にするために$\nabla_v \varepsilon = \nabla_w \varepsilon = 0$を仮定することができる）

□ To minimize the error due to truncation of the finite element mesh, infinite elements similar to those in [4] were used.
（有限要素メッシュの打切りによる誤差を最小にするために，[4]の無限要素と類似した無限要素を用いた）

□ It is possible to minimize the equivalent bandwidth.
（その等価帯域幅を最小にすることは可能である）

□ The improved algorithm can be easily applied to minimise the global roundoff noise gain of two-dimensional digital filters.
（その改良したアルゴリズムは，二次元ディジタルフィルタの大域的丸め雑音利得を最小にするのに容易に適用できる）

□ Optimization techniques have been developed which aim at minimizing the chip area for a given throughput.
（ある特定のスループットに対してチップの面積を最小にすることを目指した，最適化手法が開発された）

ポイント☞［反］maximize(最大にする)　n:minimization 最小化
☞minimum「最小・極小(の)」，minimum↔maximum(最大・極大)

model

(modeled, modeled, modeling)

〈他〉（物・事を）モデル化する

□ The propagation of electromagnetic energy inside screened rooms can be modeled by the use of various numerical techniques.
（遮蔽室内の電磁エネルギーの伝搬は，さまざまな数値手法を使用することでモデル化できる）

□ At high frequency,the p-i-n photodiode should be modeled by the transmission-

line theory.
(高周波では, p-i-nフォトダイオードは伝送線路理論によってモデル化しなければならない)

☐ The EMP source is modeled as an incident plane wave.
(EMP源は入射平面波としてモデル化される)

☐ The actuators are modelled as permanent magnetic DC motors.
(アクチュエータは永久磁石DCモータとしてモデル化する)

☐ The small-signal equivalent circuit used to model the MESFET is shown in Figure 2.
(MESFETをモデル化するのに使用した小信号等価回路を図２に示す)

☐ Fuzzy reasoning is useful for modeling the human thinking and feeling which are difficult to formulate.
(ファジィ推論は, 定式化が困難な人間の思考や感情をモデリングするのに役立つ)

☐ Therefore, origin of 1/f noise in these devices is still unknown and there are no theories modeling their noise behavior.
(それゆえ, これらのデバイスの1/f雑音の起源は今なお未知であり, この雑音の挙動をモデル化する理論は存在しない)

ポイント☞名詞としては「モデル・模型」. model(l)ing はモデル化・モデリング.
☞system model(システムモデル), stochastic model(確率的モデル)

modify

(modified, modified, modifying)

〈他〉(部分的に)修正する, 変更する, 改善する

☐ The step response of a linear control system can be modified by means of proper eignstructure assignment.

(線形制御システムのステップ応答は,適切な固有構造割当によって改善できる)

□ This circuit was slightly modified with the addition of an XOR gate to make a 9-bit pseudorandom sequence generator.
(この回路はXORゲートを加えることで少し修正し，9ビット疑似乱数列発生器を製作した)

□ In order to use the unpolarized light of gas-discharge lamps, the phase modulator has to be modified to obtain a phase shift that does not depend on the light polarization.
(ガス放電ランプの非偏光を使用するには,偏光に依存しない移相を得るために位相変調を修正しなければならない)

□ This process actually modifies the physical shape of the layout element.
(この工程は，レイアウト要素の物理的形状を実際に変更する)

□ The learning method is based on modifying the back-propagation algorithms for the purpose of on-line learning.
(その学習法は,オンライン学習を目的とした逆伝搬法アルゴリズムを修正したものに基づいている)

□ This will be done by modifying the Lyapunov functions slightly.
(これは，リャプノフ関数をわずかに変形することで行うことができる)

ポイント☞［類］correct, amend, rectify　　n:modification 修正・変更
☞「方法や性質を部分的に修正する」のニュアンスがある.

modulate

(modulated, modulated, modulating)

〈他〉（周波数・振幅などを）変調する

□ The frequency of the laser is modulated by a sinewave applied to the injection current.

（レーザの周波数は，注入電流に引加された正弦波によって変調される）

□ The first beam was modulated at 50 kHz,and the second beam was modulated at 100 kHz.
（1番目のビームは50 kHzで変調し，2番目のビームは100 kHzで変調した）

□ The frequency of the laser was directly modulated by a 1.7 Gb/s signal without the use of any preequalization circuit.
（レーザの周波数は，前置等化回路を使用することなく1.7 Gb/sの信号によって直接変調された）

□ Relative rotation of the polarizers modulates the light with a maximum light intensity produced when the polarizers are parallel and a manimum light intensity when the orientation is perpendicular.
（偏光子の相対回転は，偏光子が平行の場合に最大光度で，また方向が垂直の場合は最小光度で，光を変調する）

ポイント☞n:modulation
☞modulator は変調器．amplitude modulation(振幅変調)，frequency modulation(周波数変調)，phase modulation(位相変調)，optical modulator(光変調器)

monitor

(monitored, monitored, monitoring)

〈他〉（状態などを）監視する

□ The gas in the chamber was monitored by quadrapole mass spectroscopy.
（チャンバ中のガスは四極子型質量分析計によって監視された）

□ The phase conversion can be readily monitored by X-ray diffraction analyses.
（その位相変換はX線回折解析によって簡単に監視できる）

□ The temperature of the ambient air was monitored throughout the experiment.
（周囲の空気の温度は実験の間中監視された）

☐ A quadrupole mass spectrometer monitors the residual gas content in the chamber.
（四極子型質量分析計はチャンバ中の残留ガスの含有量を監視する）

☐ A type T copper-constantan thermocouple was used to monitor the temperature of the sample.
（T型銅－コンスタンタン熱電対は，試料の温度を監視するのに用いられた）

☐ It is useful to monitor the operating temperature of permanent magnet motors and,if possible, that of the rotor itself.
（永久磁石モータ，できれば回転子自体の動作温度を監視することは有用である）

☐ These two beams are sent along two paths of the interferometer,one of which contains the sample being monitored.
（これらの２本のビームは干渉計の二つの経路に沿って送られ，そのうちの一つの経路には監視される試料が収容されている）

ポイント☞名詞としては「監視装置・モニター」.

mount

(mounted, mounted, mounting)

〈他〉（物を）取り付ける，実装する，搭載する

☐ The specimen was mounted in the fixture between the transducers.
（試料はトランスジューサ間の取り付け器具に取り付けられた）

☐ The circuit board and power supplies are mounted in a water proof fiberglass enclosure.
（回路基板と電源は，防水ファイバグラス製筐体に取り付けられる）

☐ The coil was mounted in the bottom of a wooden box having inside length, width, and depth of 60, 40, and 20 cm, respectively.
（内側の長さ，幅，奥行きがそれぞれ60，40，20 cmの木箱の底にコイルが取

り付けられた）

□ This small portable radar can be mounted on light airplanes.
（この小型携帯型レーダは軽飛行機に搭載できる）

□ An example of a model for robot control, using an ultrasonic range sensor mounted in the robot gripper, is shown in Fig.3.
（ロボットのグリッパに取り付けられた超音波距離センサを用いた，ロボット制御のためのモデルの一例を図３に示す）

□ Devices mounted on these boards have thermal expansion properties which differ from those of the PCB.
（これらの基板上に実装されたデバイスは，PCBのとは異なる熱膨張特性をもつ）

ポイント☞n:mounting 取り付け・実装
☞surface mounting(表面実装)

move

(moved, moved, moving)

〈他〉（物を）動かす，移動させる
〈自〉（物が）動く，移動する

□ The chamber is mounted on a table which can be moved in both horizontal and vertical directions.
（そのチャンバは，水平方向と垂直方向に移動できるテーブル上に設置される）

□ The bandwidth decreases as the position of the light beam is moved further from the electrode.
（光線の位置が電極からさらに移動するにつれて，帯域幅は小さくなる）

□ A permanent magnet's south pole is moved perpendicular towards the active area of the device.

（永久磁石のＳ極は，デバイスの作用面積の方へ垂直に移動する）

□ Thus,the larger the radius is, the faster the particle moves to an equilibrium point.
（したがって，半径が大きければ大きいほど，それだけ粒子は平衡点の方へ移動する）

□ An objective lens, suspended by two parallel leaf springs, can move in a vartical direction.
（２本の並列板ばねによって吊り下げられた対物レンズは，垂直方向に動くことができる）

□ The injected hot electrons generate cold electrons and holes which then move by drift diffusion in the base.
（注入された熱い電子は，ベース中のドリフト拡散によって移動する冷たい電子と正孔を生成する）

□ As the surface moves in response to an incident ultrasonic wavefront, it causes a time-dependent phase shift to occur in the light reflected from the surface.
（表面は入射超音波頭に応じて動くため，時間に依存する移相が表面反射した光で生ずることになる）

□ As temperature increases, these curves move towards lower frequencies.
（温度が上昇するにつれて，これらの曲線は低周波の方へ移る）

□ The mobile robot is assumed to move on flat ground.
（その移動ロボットは平坦な地面上を動くものと仮定する）

ポイント☞［類］shift, migrate, transfer, transport　n:movement 動き・移動
adj:movable 移動できる，moving 動く
☞move about(動き回る)

N

need

(needed, needed, needing)

〈他〉（物・事を）必要とする

☐ High-speed circuit simulators are needed for designing large scale circuits such as LSIs and multi-chip modules.
（LSIやマルチチップモジュールといった大規模回路の設計には，高速回路シミュレータが必要である）

☐ Wideband amplifiers are needed in almost all modern microwave/lightwave communication and instrumentation systems.
（ほとんどすべての現代マイクロ波／光波通信システムと計測システムでは，広帯域増幅器が必要である）

☐ The measurement accuracy that is needed to measure the dielectric constant of anisotropic materials cannot be achieved in a routine way.
（異方性材料の誘電率の測定に必要とされる測定精度は，いつもの方法では達成できない）

☐ A high cutoff frequency at low current level is needed to reduce shot noise.
（散弾雑音を低減するには，低電流レベルにおける高遮断周波数が必要である）

☐ High performance processors need two kind of wires.
（高性能プロセッサには，２種類のワイヤが必要である）

☐ Custom layout design needs a lot of design time and cost.
（カスタムレイアウト設計には，多くの設計時間とコストが必要である）

□ The method can be used to produce the gratings needed for devices used at 1.3-μm wavelength.
(その方法は，1.3-μmの波長で使われるデバイスに必要な格子を製作するのに用いることができる)

ポイント☞［類］require
☞need to do で「～する必要がある」．名詞としては「必要性」．

neglect
(neglected, neglected, neglecting)
〈他〉（物・事を）無視する

□ $\triangle Cr$ can be neglected if a ceramic capacitor with zero temperature coefficient is chosen.
(もし温度係数がゼロの磁器コンデンサを選べば，$\triangle Cr$は無視できる)

□ The curve in Fig.4 represents the behavior of the lines under the assumption that the skin effect can be neglected.
(図４の曲線は，表皮効果を無視できるという仮定のもとでの線の挙動を表している)

□ We can neglect the effect on the input signal transfer function since it is very small.
(入力信号の伝達関数への影響は非常に小さいため，無視できる)

□ We have neglected the effects of the SQUID on the input circuit.
(ＳＱＵＩＤが入力回路へ及ぼす影響は無視した)

□ Neglecting V_F in eqn.2, it can be seen that the value of the model control voltage V_T is the averaged value of a rectangular waveform with a duty ratio of 2D.
(方程式２のV_Fを無視すれば，モデル制御電圧V_Tの値は，衝撃係数が２Ｄの矩形波形の平均値となることが理解できる)

ポイント☞［類］ignore,disregard　n:negligence 怠慢・不注意
adj:neglectful,negligent 怠慢な
☞「十分な注意を払わないで無視する」のニュアンスがある．

note

(noted, noted, noting)

〈他〉（物・事に）注意する，注目する，気づく，言及する
note that節：～ということに気づく，注目する

□ It should be noted that no direct adaptation of the parameter vector $\Theta(t)$ is performed, since this vector does not affect the output error $\nu(t)$.
(パラメータベクトル$\Theta(t)$の直接適応は行われないということに注目しなければならない。というのは，このベクトルは出力誤差$\nu(t)$に影響を与えないからである)

□ The use of these contact potentials to measure the composition of gaseous atmospheres has been noted in this literature.
(気体雰囲気の組成を測定するために，これらの接触電位差を使用するということは，この文献に言及されている)

□ We noted that another main problem was to solve this partial differential equation.
(もう一つの主な問題は，この偏微分方程式を解くことであるということに気づいた)

□ It is worth noting that the offset can be adjusted easily by means of a resistor.
(オフセットは抵抗器によって容易に調節できるということは注目する価値がある)

ポイント☞［類］notice, pay attention to　adj:noted 著名な
☞名詞としては「メモ・注目」．

O

observe

(observed, observed, observing)

〈他〉（物・事を）観測する，観察する，認める

observe that節：〜ということに気づく

☐ Resonance phenomenon is observed at a frequency around 9 MHz.
（共振現象は9 MHz付近の周波数において観測される）

☐ No hysteresis effects were observed in any of the PI curves that showed the sharp turn-on.
（鋭いターンオンを示したPI曲線のどれからもヒステリシスの影響は観測されなかった）

☐ It can be observed that the computed voltages are smaller than measured values.
（計算された電圧は測定値よりも小さいということが認められる）

☐ We observe that the error decreases as δ is increased from zero.
（δが零から増加するにつれて誤差は減少するということに気づく）

☐ From Theorem 2, we observe that this statement is not correct.
（定理2より，この命題は正しくないということに気づく）

☐ This paper investigates various methods for processing these signals to observe the velocity in real time.
（実時間で速度を観測するために，これらの信号を処理するさまざまな方法について調べる）

ポイント☞n:observation 観測・観察　　adj:observant 観察の鋭い

☞observatory は観測所．weather observation(気象観測)

obtain

(obtained, obtained, obtaining)

〈他〉（物・事を)得る，獲得する

☐ Equation (3.2) is obtained from an one-dimensional model.
(方程式(3.2)は一次元モデルから得られる)

☐ Sinusoidal outputs are usually obtained by nonlinear wave shaping of constant amplitude triangular waves.
(正弦波出力は，通常は一定振幅の三角波を非線形波形整形することによって得られる)

☐ Designers can rapidly obtain required layouts using these tools.
(設計者はこれらのツールを用いて，必要とされるレイアウトを速やかに得ることができる)

☐ Logic simulations can be used to obtain further information about problems.
(論理シミュレーションは，問題についてのさらに進んだ情報を獲得するのに用いることができる)

☐ The obtained mutual inductance was 30 pH and was independent of the temperature from 4.8 K to 12.4 K.
(得られた相互インダクタンスは30 pHで，4.8 K～12.4 Kの温度とは無関係だった)

ポイント☞［反］lose(失う)　［類］acquire,attain,get,gain　adj:obtainable 入手可能な

☞「努力することで何かを得る」のニュアンスがある．

occur

(occurred, occurred, occurring)

〈自〉（物・事が)起こる，生じる，現れる，存在する

☐ Strong excitation of surface waves occurs at the third-order harmonic frequency.
（表面波の強い励振は三次高調波振動数で生じる）

☐ In the case of ac magnetic fields,the important effects occur with induced dipole moments.
（交流磁場の場合，重要な効果が誘導双極子モーメントによって生ずる）

☐ Time delays occure frequently in process control problems.
（時間遅れはプロセス制御問題でしばしば現れる）

☐ The same problem occurs for isotropic media.
（等方性媒質の場合，同じ問題が生じる）

☐ These problems occur in both digital and analog systems.
（これらの問題はディジタルシステムとアナログシステムの両方に現れる）

ポイント☞［類］arise, happen, take place, give rise to, result from　　n:occurrence 発生・出来事

☞「あることが特定の時期に起こる」のニュアンスがある．

offer

(offered, offered, offering)

〈他〉（物・事を）提供する，与える，示す

☐ This paper offers an effective approach for calculating these parameters by the finite element method.
（本論文では，これらのパラメータを有限要素法によって計算する効果的なアプローチを提供する）

☐ The results of over-voltage tests offer some insight into possible behaviors of heaters under normal testing conditions.
（その過電圧試験の結果は，通常の試験時でヒータの起こりうる挙動への見通しを与える）

☐ Intergrated optical devices <u>offer</u> a number of attractive features for use in fiber optic gyros.
(光集積デバイスは光ファイバジャイロ用として，多くの魅力的な特徴を有している)

☐ Such optical interactions within fibers <u>offer</u> intriguing capabilities in a variety of optical devices such as optical amplifiers, lasers, wavelength convertors, soliton transmitters, and passive devices.
(ファイバ内のそのような光相互作用は，光増幅器，レーザ，波長変換器，ソリトン送信機，受動装置といったさまざまな光デバイスにおいて，魅力的な能力をもたらす)

ポイント☞［類］provide,furnish　　n:offering 提供・製品
☞名詞としては「提供・申し出」.

---

operate
(operated, operated, operating)

〈他〉（機械・装置などを）動作させる，操作する，運転する
〈自〉（機械・装置などが）動作する，作動する，機能する，動く

☐ The converter can be <u>operated</u> as a resonant converter.
(その変換器は共振型変換器として動作する)

☐ A number of groups have already <u>operated</u> SQUIDs successfully.
(多くのグループがすでにSQUIDを動作させるのに成功している)

☐ Superconductive alloys of lead and niobium <u>operate</u> at temperatures only a few degrees above absolute zero.
(鉛とニオブの超伝導合金は，絶対零度よりほんの数度高い温度で機能する)

☐ At this high electric field, the MQW can <u>operate</u> as an optical detector.
(この高電場において，MQWは光検出器として動作する)

☐ In order to <u>operate</u> at the high frequency the switching losses should be

decreased sufficiently.
(高周波での動作のためには,スイッチング損失を十分に減少させなければならない)

☐ This shows that these oscillators are operating in the second harmonic mode.
(このことは,これらの発振器が第二高調波モードで動作しているということを示している)

☐ Low-power, high-density memory is indispensable for high-performance battery operated equipment such as notebook PC's.
(低電力,高密度メモリは,ノートブック型PCのような高性能バッテリー運用の装置には欠かせない)

☐ A semiconductor laser operating at 830 nm was used as the light source.
(830 nmで作動する半導体レーザが光源として用いられた)

ポイント☞ [類]run, work, drive, function, manipulate　n:operation 操作・演算　adj:operational 運転できる
☞operator はオペレータ・操作員．Boolean operation(ブール演算), operational amplifier(演算増幅器)

optimize

(optimized, optimized, optimizing)

〈他〉(物・事を)最適化する

☐ The delay line length was optimized for an operating frequency of 1 GHz.
(1 GHzの動作周波数のために,遅延線の長さを最適化した)

☐ Consequently, brightness and resolution cannot be optimized independently.
(したがって,輝度と解像度は別々に最適化できない)

☐ The receiver in this model has not been optimized for the impulsive/ Gaussian noise.
(そのモデルにおける受信機は,インパルス／ガウス雑音に対して最適化され

なかった)

☐ This choice has to be made so as to optimize the system behaviour.
(システムの動作を最適化するために，この選択が行われなければならない)

☐ Figure 3 illustrates the silicon developed to optimize the on-resistance R_{DC} and breakdown voltage V_{DC} of the power DMOS transistors.
(図3は,パワーDMOSトランジスタのオン抵抗R_{DC}と破壊電圧V_{DC}を最適化するために開発されたシリコンを示している)

☐ For applications such as phase sensors,phase modulators are useful for optimizing sensor performance.
(位相センサのような用途の場合,位相変調器はセンサの性能を最適化するのに役立つ)

ポイント☞n:optimization 最適化
☞optimization problem(最適化問題), constrained optimization(条件つき最適化)

overcome
(overcame, overcome, overcoming)
〈他〉(障害・困難・問題などに)打ち勝つ，克服する

☐ Many of the disadvantages of dc motors can now be overcome by the very compact semiconductor control amplifiers.
(直流電動機の短所の多くは，非常にコンパクトな半導体制御増幅器によって，今や克服できる)

☐ In this paper we report the results of a new technique which overcomes this problem.
(本論文では，この問題を解決する新手法の成果について報告する)

☐ To overcome these difficulties, a new method has been proposed in Reference [2].

（これらの困難な点を克服するために，参考文献[2]では新しい方法が提案されている）

□ The actuator is driven sufficiently hard by each pulse to overcome the static friction.
（静止摩擦に打ち勝つために，アクチュエータは各パルスによって十分強力に駆動される）

ポイント☞「困難を排して障害に打ち勝つ」のニュアンスがある．

p

pass

(passed, passed, passing)

〈他〉（物を）通過させる，通す
pass A through B：AをBの中を通過させる
〈自〉（物が）通過する，通る
pass through A：Aの中を通過する

□ The signal is passed through the 60 Hz notch filter having a bandwidth of 10 Hz.
（信号は10 Hzの帯域幅を有する60 Hzノッチフィルタを通過する）

□ The pulses are passed through a low-pass filter to eliminate noise.
（雑音を除去するためにパルスは低域フィルタを通過する）

□ Recived signals were passed into a preamplifier.
（受信信号は前置増幅器に送られた）

□ The diffracted light can pass between these stripes toward the projection screen.
（回折光は投影スクリーンに向かってストライプ間を通過する）

☐ The sum of magnetic flux Φ(t) passing through a closed loop of the secondary coil can be evaluated by dividing the primary coil into a number of straight-line segnents and calculating the magnetic field contributed by each segment using Eq.(1).
(二次コイルの閉ループを通過する磁束Φ(*t*)の総量は,一次コイルを多くの直線分に分割し,方程式(1)を用いて各線分がもたらす磁場を計算することで求められる)

☐ Venturi meters measure velocity by calculating the difference in pressure as flow passes through a restricted area.
(ベンチュリ計は,流れが限られた面積を通過する時の圧力差を計算することで,速度を測定する)

ポイント☞n:passage 通過・通行　　adj:passable 通れる

perform
(performed, performed, performing)
〈他〉（仕事・試験などを)行う，する
〈自〉（機械などが)機能する，動作する

☐ Creep tests were performed inside a heated or cooled copper cylinder.
(過熱された，あるいは冷却された銅の円柱内部でクリープ試験が行われた)

☐ A series of tests have been performed on the fiber optic torquemeter.
(一連の試験を光ファイバトルク計を用いて実施した)

☐ When fast frequency switching is not essential, tuning can be performed with an optical encoder.
(高速周波数スイッチングが重要でなければ,同調は光エンコーダを用いて行うことができる)

☐ The tracking controller discussed in this paper performs only feedforward control.
(本論文で議論した追従制御器は，フィードフォワード制御のみ実行する)

□ Traditional circuit simulators like SPICE and ASTAP become inefficient when they perform the transient analysis of large integrated circuits.
(SPICEやASTAPといった従来の回路シミュレータは，大規模集積回路の過渡解析を実施する場合には能率が上がらない)

□ The circuit performs very well at high frequencies and exhibits significant improvements in terms of temperature stability.
(この回路は高周波で非常によく動作し，温度安定性の点で著しい向上を示す)

ポイント☞［類］do, conduct, run, execute　　n:performance 実行・性能
☞「一定の手順に従って行う」のニュアンスがある．

pick
(picked, picked, picking)
〈他〉（物・事を）選ぶ，選択する
pick up A：Aを拾いあげる，集める

□ The voltage waveform at the output node was picked up by the FET probe with 2pF input capacitance.
(出力ノードでの電圧波形は，入力容量が2pFのFETプローブによって集めた)

□ Fading is caused if the receiving antennas, like the very small antennas used in mobile radio links, pick up multipath reflections.
(移動無線リンクで使用される超小型アンテナのような受信アンテナが多重反射を拾いあげるならば，フェージングを引き起こす)

□ To compare the two methods described, we picked four MODFET's with dc characteristics typical of the particular wafer (C195).
(述べられた二つの方法を比較するために，特定のウェハ(C195)に特有な直流特性を有する4個のMODFETを選んだ)

□ The electrostatic noise picked up by the electrodes contains mainly low-frequency components.

（電極によって拾われた静電雑音は，主に低周波成分を含んでいる）

ポイント ☞［類］choose, select, prefer
☞「複数の物より慎重に判断することによって選ぶ」のニュアンスがある．名詞としては「選択」．pickup はピックアップ．

place
(placed, placed, placing)
〈他〉（物を）置く，配置する，設置する

☐ The breadboard circuit was placed on a table 80 cm above a ground plane.
（ブレッドボード回路はテーブル上に接地面より80 cm高く置かれた）

☐ The eigenvalues of the closed-loop system matrix(G－HK) can be placed in any specified locations if and only if the pair(G,H) is controllable.
（閉ループシステム行列(G－HK)の固有値は，組(G,H)が可制御の時，かつその時に限って指定された位置に配置される）

☐ The sensor is placed in contact with the sample,and its complex reflection coefficient is measured.
（センサは試料と接触して置かれ，その試料の複素反射係数が測定される）

☐ The device was placed in semianechoic chamber that is regularly used for EMC qualification testing.
（装置は，ＥＭＣ認定試験のために定期的に使用される半無響室に設置された）

☐ We placed the silicon wafer on an electrically controlled heating plate.
（シリコンウェハを電気的に制御された加熱板上に置いた）

ポイント ☞［類］put, locate, install　　n:placement 配置
☞名詞としては「場所・位」．in place of(～の代わりに)，in the first place(まず第一に)，decimal place(小数位)

play

(played, played, playing)

〈他〉 (～の役割を)果す

play～role[part] in A：Aにおいて～の役割を果す

☐ Coherent detection techniques play an important role in the development of optical fiber sensors.

(同期検波手法は，光ファイバセンサの開発において重要な役割を果す)

☐ In this connection the dynamic range of the apparatus plays a fundamental role since it determines the maximum scan angle.

(この接続の場合，装置のダイナミックレンジは重要な役割を果す。というのは，それが最大走査角を決めるからである)

☐ Energy functions play an essential role in direct methods for the analysis of power system transient stability.

(電力系統の過渡安定度解析のための直接法において，エネルギー関数は重要な役割を果す)

ポイント☞playback は再生，CD player はCDプレーヤー．

plot

(plotted, plotted, plotting)

〈他〉 (図形などを)プロットする

plot A against B：AをBに対してプロットする

☐ The simulation results are plotted in Fig.4.

(シミュレーション結果は図４にプロットしてある)

☐ The normalized gain margin of the main modes is plotted against the normalized detuning coefficient in Figure 3.

(主モードの正規化された利得余裕は，図３に正規化された離調係数に対してプロットしてある)

☐ In Fig.5,We plot the current－voltage characteristics of the circuit.
（図5に回路の電流・電圧特性をプロットする）

☐ Plotting the difference in ratio form as shown in Fig.5 illustrates this relationship more clearly.
（図5に示すように，比率の形の違いをプロットすることで，この関係はより明確に図示できる）

ポイント☞名詞としては「プロット」．plotter はプロッタ・作図装置．

polarize
(polarized, polarized, polarizing)
〈他〉（光を）偏光させる

☐ Light is polarized by mounting a fixed polarizer in the light path.
（光路に固定偏光子を取り付けることで偏光する）

☐ Except for thin films in the 0-100 nm range,the incident light should be linearly polarized perpendicular to the plane of incidence to achieve higher sensitivity.
（0～100 nmの薄膜を除けば，高感度を達成するために入射光は入射面に垂直方向に直線偏光しなければならない）

☐ For light which is polarized perpendicular to the molecular orientation the refractive index is voltage independent.
（分子配向に垂直に偏光した光の場合，屈折率は電圧には依存しない）

☐ The calculation shows that incident light polarized in the x-direction has x-and y-components when it is reflected and is therefore elliptically polarized.
（X方向に偏光した入射光はそれが反射し，その結果楕円偏光する場合に，x成分とy成分を含むということを，計算により示す）

ポイント☞n:polarization 偏光
☞linear polarization(直線偏光)

predict

(predicted, predicted, predicting)

〈他〉（事を）予測する

predict that節：～ということを予測する

☐ The behavior can be accurately predicted using mode coupling theory.
（その挙動はモード結合理論を用いて正確に予測できる）

☐ The model predicted the performance of a real system with good accuracy for piston displacement responses.
（そのモデルによって，行程容積の応答に対する実システムの性能を正確に予測した）

☐ Although this problem cannot be solved analytically,we can predict some general characteristics.
（この問題は解析的に解けないが，大まかな特性は予測できる）

☐ Computer simulations predict that the circuit of Fig.3 can operate up to 25 GHz if the jjs have $j_c = 2$ kA/cm^2.
（コンピュータシミュレーションは，図３の回路はjjsが$j_c = 2$ kA/cm^{2}ならば，25 GHzまで動作するということを予測している）

☐ We attempted to predict the drain voltage dependence of the power spectral density.
（パワースペクトル密度のドレイン電圧への依存性を予測することを試みた）

☐ Since the capacitance is not yet known for high-temperature superconducting tunnel junctions, it is impossible to predict the effect on switching speed for these circuits.
（高温超伝導トンネル接合で，この静電容量はまだわかっていないので，これらの回路のスイッチング速度への影響を予測することは不可能である）

☐ The measurements of temperature were compared with those predicted by the model.

(温度の測定値をモデルで予測した測定値と比較した)

□ We present a method for predicting the electron holographic interference pattern from a given magnetization distribution.
(ある特定の磁化分布から電子線ホログラフィー干渉パターンを予測する方法を示す)

□ Predicting the cycle time of a processor in which performance is the overriding concern is actually becoming simpler as technology advances.
(性能がすべてに優先する関心事であるプロセッサのサイクル時間を予測することは，技術が進歩するにつれて実際に容易になってきている)

ポイント☞［類］anticipate, expect, forecast　　n:prediction 予測
adj:predictable 予測できる
☞「事実・データなどに基づいて予測する」のニュアンスがある.

prepare
(prepared, prepared, preparing)

〈他〉（物・事を）用意する
（試料などを）作製する
prepare A for B：AをBのために用意する

□ A 1.3 μ m InGaAsP two-electrode DFB laser which was grown by liquid phase epitaxy (LPE) was prepared for the experiment.
(液相エピタキシー（LPE）によって成長させた 1.3 μ m InGaAsP 2電極DFBレーザを，実験のために用意した)

□ Specimens were prepared by melting the alloys in a nitrogen atmosphere.
(試料は窒素雰囲気中で合金を溶融することで作製された)

□ These alloys can be prepared in the form of an amorphous ribbon.
(これらの合金は，非晶質リボンの形に作製できる)

□ The YBaCuO polycrystalline thin film was prepared using this process.

(YBaCuO多結晶薄膜は，この工程を用いて作製された)

ポイント☞［類］ build, make, manufacture, fabricate, assemble, form, produce, construct　n:preparation用意　adj:preparatory準備の

present

(presented, presented, presenting)

〈他〉（論文・報告書などを)提出する，示す

☐ A new algorithm is presented for adaptive parameter estimation for a low-pass Butterworth system model.

(低域通過バターワースシステムモデル用の適応パラメータ推定のために,新しいアルゴリズムを示す)

☐ The theory of measurement of trapping centers parameters using the admittance spectroscopy method has been presented elsewhere.

(アドミッタンス分光法を用いる捕獲中心のパラメータ測定理論は他のところで提出した)

☐ In this paper, we first present a state-space representation of a matching network.

(本論文では，まず初めに整合回路網の状態空間表現を与える)

☐ This paper presents a simple way to use PVDF piezoelectric films to construct a multiple active element ultrasonic transducer.

(本論文では，多能動素子超音波変換器を組み立てるために,PVDF圧電膜を使用する簡単な方法を提出する)

ポイント☞n:presentation 提出・プレゼンテーション

presume

(presumed, presumed, presuming)

〈他〉（事を)推定する，仮定する

presume that節：～だと推定する

presume A (to be) B：AがBだと推定する

□ It is presumed that no magnetic field arises in the conductor, when the electric current flows uniformly in it.
(電流が導体中を一様に流れる場合,その導体には磁場が生じないということが推定される)

□ The mean-field approximation is presumed to be valid.
(その平均場近似は有効であると推定される)

□ We presume that appropriate alloy grading techniques would allow these contact resistances to be obtained.
(適切な合金グレーディング手法によって,これらの接触抵抗が求められるものと推定している)

ポイント☞[類] assume,suppose,hypothesize　n:presumption推定・仮定　adj:presumable 仮定できる・ありそうな　adv:presumablyたぶん
☞「確証はないが確信をもって推定する」のニュアンスがある."目的語＋to do"は不可. "presume to do"だと,「あえて～する」の意味となる.

prevent

(prevented, prevented, preventing)

〈他〉(物・事を)妨げる,防ぐ
prevent A from doing：Aが～するのを防ぐ

□ Circuit complexity,IC defect anomalies,and economic considerations prevent complete validation of VLSI circuits.
(回路の複雑さ,IC欠陥異常,経済上の考慮事項により,VLSI回路の完全な妥当性検査は妨げられる)

□ This array prevents water from forming films on the die surface.
(この配列をすることで,ダイ表面に水の膜が形成されるのを防ぐ)

□ The choke prevents the return current from flowing up the transmission line to

the body surface.
（そのチョークは伝送線路から体表面へ帰還電流が流れるのを防ぐ）

☐ The high-speed buffer amplifier is used as a driver, which prevents the load from affecting the proper operation of the active network.
（高速緩衝増幅器はドライバとして使用され，負荷が能動回路網の正常な動作に影響を及ぼすのを防ぐ）

☐ Shielding is a technique used to control noise by preventing transmission of EMI from the noise source to the receiver.
（遮蔽は，雑音源から受信機へＥＭＩを伝えないようにすることで，雑音を抑制するのに用いられる一つの手法である）

ポイント☞［類］hinder, obstruct　n:prevention 防止・予防　adj:preventive 防止の・予防の
☞「あらかじめ予防手段を講じることで，不具合なことの発生を不可能にする」のニュアンスがある．無生物主語の構文で多用される．

process

(processed, processed, processing)

〈他〉（信号・データなどを）処理する

☐ Data-acqusition and sampled-data control algorithms must be processed at regular time intervals.
（データ収集アルゴリズムとサンプル値制御アルゴリズムは，一定の時間間隔で処理しなければならない）

☐ A high resolution macrowave imaging system must process large amounts of data.
（高解像度マイクロ波撮像システムは，大量のデータを処理しなければならない）

☐ Kalman filtering is used to process signals which are corrupted by noise.

(カルマンフィルタリングは，雑音で汚された信号を処理するのに用いられる)

☐ Counters are the most common systems for processing phase-Doppler-difference signals.
(カウンタは位相ドップラー差信号を処理するための，最も一般的なシステムである)

ポイント☞ [類] treat
☞名詞としては「処理・過程・プロセス」. processing も処理. process control(プロセス制御)，in the process(進行中で)，data processing(データ処理)，image processing(画像処理)，parallel parallel(並列処理)

produce
(produced, produced, producing)
〈他〉（物・事を）作り出す，生み出す，発生させる，引き起こす

☐ Both position and velocity information can be produced simultaneously from the same transducer.
(位置情報と速度情報は，同じトランスジューサから同時に発生させることができる)

☐ An NMR spectrum is produced by holding the magnetic field constant and sweeping the RF frequency.
(NMRスペクトルは，磁場を一定に保持し，ＲＦ周波数を掃引させることによって生じる)

☐ Most band-pass transformations produce symmetrical transfer functions only.
(たいていの帯域通過変換は対称伝達関数のみを生成する)

☐ Unfortunately variations in both environmental conditions and material composition can also produce changes in the optical phase.
(残念ながら，環境状況と材料組成の変化は光学位相の変化も引き起こす)

☐ External Fabry-Perot interferometers can be used to produce single frequency emission.
(外部ファブリーペロ干渉計は，単一周波数放射を発生させるのに使用できる)

ポイント☞［類］ build, make, manufacture, fabricate, assemble, form, prepare, construct, generate, develop　n:product製品・生産物，production 製造・生産　adj:productive 生産的な
☞「材料から製品などを大量に製造する」のニュアンスがある．
production line(生産ライン)，production control(生産管理)

propagate

(propagated, propagated, propagating)

〈他〉 (音・光・電磁波・熱などを)伝える，伝播させる，伝搬させる
〈自〉 (音・光・電磁波・熱などが)伝わる，伝播する，伝搬する
propagate through A：Aの中を伝わる

☐ These waves are propagated without loss along an interface between two solid media.
(これらの波は，二つの固体媒体間の界面に沿って損失なしに伝搬する)

☐ There is the possibility that the TE_{11} mode may propagate in the waveguide at the frequency of the TE_{01} resonance.
(このTE_{11}モードはTE_{01}の共振周波数で導波管を伝搬する可能性がある)

☐ When these waves propagate through the horn type beamwidth compressor, the elastic nonlinearity is to be enhanced by compression ratio.
(これらの波がホーン型ビーム幅圧縮器中を伝搬する時には，弾性非線形性を圧縮比だけ強めるべきである)

☐ The experiment has confirmed the insignificance of signals propagating in commercial laminated epoxy-glass PCB material.
(市販の積層エポキシ・ガラスPCB材料を伝搬する信号は重要ではないということを，その実験で裏づけた)

ポイント☞n:propagation 伝播・伝搬
☞ionospheric propagation(電離層伝搬), propagation delay(伝搬遅延), propagation constant(伝搬定数)

propose
(proposed, proposed, proposing)

〈他〉（事を）提案する
propose to do：～するつもりである

□ Optical methods have been proposed for both flow and level measurement but are only applicable in a few selected situations.
（流量および水準測定のために，光学的方法が提案されているが，少数の選ばれた状況でしか適用できない）

□ Many fast algorithms have been proposed for computing the discrete Fourier transformation.
（離散フーリエ変換を計算するために，多くの高速アルゴリズムが提案されてきた）

□ Two types of parallel A/D converter have been proposed using Josephson junction technology.
（ジョセフソン接合技術を用いた二種類の並列A/D変換器が提案された）

□ This paper proposes a procedure to solve these problems.
（本論文では，これらの問題を解くための手順を提案する）

□ We propose to apply the method described above for the detection of non-stationary signals.
（非定常信号の検出に上記の方法を適用するつもりである）

ポイント☞［類］suggest, put forward　n:proposal 提案，proposition 提案
☞that節と伴うことも可能だが，その場合は that節内の動詞は仮定法現在か should～となる．

protect

(protected, protected, protecting)

〈他〉（損傷・危険などから）守る，保護する

protect A from B：AをBから守る

☐ This transformer is protected from overvoltages by a silicon carbide varistor.
（この変圧器は炭化けい素バリスタによって過電圧から守られている）

☐ The computer and oscilloscope were protected from transients on the mains by a passive low-pass filter.
（コンピュータとオシロスコープは，受動低域フィルタによって送電線の過渡現象から保護された）

☐ In some cases, specially-designed barriers must be installed inside the enclosure to assure that the input devices are protected from overvoltage or overcurrent conditions.
（場合により，入力装置が過電圧や過電流から保護されるという保証が得られるように，特別に設計された障壁をエンクロージャ内に設置しなければならない）

☐ The shields in the superconducting synchronous generator protect the superconducting field winding from varying fields and damp the generator oscillations.
（超伝導同期発電機におけるこのシールドは，変動磁場から超伝導界磁巻線を保護し，発電機の振動を抑える）

ポイント ☞n:protection 保護　adj:protective 保護の

☞data protection(データ保護)，memory protection(記憶保護)，electric protective device(電気保護装置)

prove

(proved, proven, proving)

〈他〉（事を）証明する，立証する

prove A (to be) B：AがBであると証明する

	prove that節：～ということを証明する
〈自〉	（事が）わかる，判明する
	prove（to be）A：Aであるとわかる，判明する

□ Equation (36) will be proved by calculating $y_j(x,t)$.
（方程式(36)は$y_j(x,t)$を計算することで証明できる）

□ We first prove some properties of the optimal cost function in Lemma 1.
（最初に補助定理１中の最適費用関数のいくつかの性質を証明する）

□ In the course of proving Theorems 2 and 3,we proved that $f_1(t)/g_1(t)$ and $f_2(t)/g_2(t)$ approach some finite nonzero constants.
（定理２と３の証明の過程で，$f_1(t)/g_1(t)$と$f_2(t)/g_2(t)$は有限な零でない定数に近づくということを証明した）

□ The introduction of this adaptation mechanism allows us to prove that the tracking error asymptotically converges to zero and that all the trajectories of the system remain bounded.
（その適応機構の導入により，トラッキング誤差は漸近的に零に収束し，システムの軌道はすべて有界のままであるということを証明できる）

□ These properties will be used in proving the theorem given in §2.1 and in developing the algorithm given in the next section.
（これらの性質は，§2.1で与えられた定理を証明する際，およびその次の節で与えられるアルゴリズムを開発する際に用いられる）

□ The experiments proved to be rather difficult and thus required a novel approach to data collection.
（その実験はかなり難しく，したがってデータ収集には斬新なアプローチが必要であることがわかった）

□ The use of artificial neural networks in system identification, fault detection and fault diagnosis problems proves to be particularly advantageous for all applications that require a high operating speed.
（システム同定，故障検出，故障診断問題に人工ニューラルネットワークを利

用することは，高動作速度が必要な用途すべてに対してとりわけ都合がよいことがわかる）

☐ This technique has proven to be very useful in the development and analysis of video display terminals.
（この手法は，ビデオ表示端末装置の開発および解析で非常に有用であることがわかっている）

☐ These methods may prove beneficial in the detection and diagnosis of component malfunctions in the system.
（これらの方法は，システムの構成要素の誤動作の検出と診断において有益であることがわかる）

☐ The generalization of SISO technique has proven to be a useful approach to the design of multivariable feedback control systems.
（そのＳＩＳＯ手法の一般化は，多変数フィードバック制御システムの設計に有用なアプローチであることがわかっている）

☐ Optical heterodyne technique has proved to be a powerful method for the detection of weak optical signals.
（光ヘテロダイン手法は，弱い光信号の検出のための強力な方法であることがわかっている）

ポイント☞n:proof 証明
☞alternative proof(別証明)

provide

(provided, provided, providing)

〈他〉（物を）供給する，与える，備える
provide A with B：AにBを供給する
provide A for B：AをBに供給する

☐ The electrical insulation of the coils from the steel of the yokes is provided by having a sufficiently large air gap.

(十分に大きなエアギャップを設けることで,コイルを継鉄の鋼から絶縁できる)

☐ The LCD can be provided with colour filters.
(そのLCDにはカラーフィルタを備え付けることができる)

☐ These units provide a speaker dependent voice recognition system.
(これらの装置を用いれば,話者依存音声認識システムを実現することができる)

☐ The location of transfer function poles and zeros provides valuable information for circuit designers.
(伝達関数の極と零点の位置により，回路の設計者は貴重な情報が得られる)

☐ Control theory provids methods to deal with such periodic disturbances.
(制御理論により，そのような周期的外乱を扱うための方法が得られる)

☐ The analyzer provides an inexpensive and accurate means of examinig the harmonic content of a electric power system.
(そのアナライザは,電力系統の高調波成分を検査する安価で正確な方法を与える)

☐ The oscillator provides very pure sine waves and does not suffer from startup problems.
(その発振器は非常に純粋な正弦波を生成し，始動の問題を抱えない)

☐ The LDV system can provide spatial resolution from 20 to 100 μm which cannot be obtained with any other method.
(そのＬＤＶシステムでは,他の方法では得られない20～100 μmの空間分解能を得ることができる)

☐ The above lemma provides a means of designing I/O robust control.
(上記の補助定理により，Ｉ／Ｏロバスト制御設計の方法が得られる)

ポイント☞［類］supply, serve, furnish, feed　n:provision 供給

☞「前もって準備して必要な物や有益な物を供給する」のニュアンスがある.

Q

quantize

(quantized, quantized, quantizing)

〈他〉（信号・エネルギーなどを）量子化する

□ No matter how fast the signal is quantized, if the ADC cannot sample the signal fast enough, quantization speed is wasted.

（ＡＤＣが十分速く信号を標本化できないならば, たとえ信号がどんなに速く量子化されるにしても，量子化速度には無駄が生じる）

□ This circuit quantizes the input signal in the same way as in a conventional delta-sigma modulator.

（この回路は, 従来のデルターシグマ変調器と同じ方法で入力信号を量子化する）

□ After quantizing the analog signal, the result must be encoded into an *N*-bit binary form.

（アナログ信号を量子化した後に, その結果を*N*ビットの２進形に符号化しなければならない）

ポイント☞n:quantization 量子化

☞quantizerは量子化器．quantization noise(量子化雑音)，vector quantization(ベクトル量子化)

R

radiate

(radiated, radiated, radiating)

〈他〉（電磁波・光・熱などを）放射する

□ The microstrip antennas used radiate a perfectly linear polarized wave because of their excitation on the TM_{10} mode.
（使用されるマイクロストリップアンテナはTM_{10}モードで励振するため，完全な直線偏波を放射する）

□ Every object whose temperature is above absolute zero radiates energy at a rate which is proportional to the fourth power of its absolute temperature.
（絶対零度以上の温度をもつあらゆる物体は，絶対温度の 4 乗に比例してエネルギーを放射する）

□ To measure amplitude and phase of electric field radiated from spherical dipole antennas, a vector voltmeter is utilized.
（球面ダイポールアンテナから放射される電界の振幅と位相を測定するのに，ベクトル電圧計を利用した）

□ The radiated electric field is obtained from $H(r)$ using Maxwell's equation $E(r)$.
（放射電界はマクスウェル方程式$E(r)$を用いて$H(r)$から得られる）

□ In the horizon plane the radiated electric field is usually zero unless the folloing condition holds:
（水平面の場合，放射電界は次の条件が成り立たない限り，通常は零である）

ポイント☞［類］emit　n:radiation 放射　adj:radiative 放射性の
☞形容詞 radioactive も「放射性の」. electromagnetic radiation(電磁放

射), radiation dose(放射線量), radioactive waste(放射性廃棄物)

range

(ranged, ranged, ranging)

〈自〉(範囲・程度などが)及ぶ，わたる

range from A to B：AからBに及ぶ

☐ For typical EMP bursts, the Compton electron source region ranges from 30 to 50 km above the earth.

(典型的なＥＭＰバーストの場合，コンプトン電子源は地上30～50 kmの範囲に及ぶ)

☐ In aircraft, for example, structural shielding effects range from 20 to 100 dB.

(例えば航空機の場合，構造遮蔽効果は20～100 dBの範囲にわたる)

☐ Systems based on artificial neural networks (ANNs) are widespread today in various fields of application ranging from pattern recognition to process control.

(人工ニューラルネットワーク(ANN)に基づくシステムは，今日ではパターン認識からプロセス制御にわたる，さまざまな応用分野に普及している)

ポイント☞進行形は不可．名詞としては「範囲」．

reach

(reached, reached, reaching)

〈他〉(ある状態・結果などに)達する

☐ This procedure is repeated until the final operating point is reached.

(この手順は，最終動作点に到達するまで繰り返される)

☐ When the temperature of the defect reaches the critical temperature, breakdown occurs.

(欠陥の温度が臨界温度に達すると，破損が起こる)

☐ Another difficulty related with ferromagnetic materials is the abrupt variation in the physical properties when the temperature reaches the Curie point.
(強磁性材料に関係するもう一つの困難な点は,温度がキュリー点に達する時に物理的性質が突然変化することである)

☐ Once the electron velocity reaches the saturation velocity, the drain current is described by (1).
(電子速度が飽和速度に達すると，ドレイン電流は(1)で書き表される)

☐ The SAR distribution reaches its highest value of 280 W/kg per watt near the tip located at the 40-mm mark along the axial direction.
(ＳＡＲ分布は，軸方向に沿った40 mm のマークに位置する先端近くで，ワット当たり最高280 W/kgの値に達する)

ポイント☞自動詞用法もあるが，多くは他動詞で用いられる．この場合，前置詞は必要ない．名詞としては「及ぶ範囲」.

react
(reacted, reacted, reacting)
〈自〉（物・事が）反応する，対応する
react with A：Aと（化学）反応する

☐ When any material enters the sensing field of a capacitive sensor, the sensor can react.
(容量型センサの検出領域に材料が入ると，センサは反応する)

☐ The robots used in flexible manufacturing processes must adapt and react to rapidly changing work environments.
(フレキシブル製造工程で使われるロボットは,急速に変化する作業環境に順応し，対応しなければならない)

☐ This trigger mechanism has to react to significant changes in the sine waveform.
(このトリガー機構は，正弦波形の著しい変化に反応しなければならない)

☐ This amine reacts with the resist in the exposed regions.
（このアミンは露光領域においてレジストと反応する）

ポイント☞［類］respond　　n:reaction 反応
☞chemical reaction(化学反応)

realize
(realized, realized, realizing)
〈他〉（物・事を）実現する

☐ A bandpass filter have been realized on a 1.5 mm^2 die.
（帯域通過フィルタは1.5 mm^2のダイ上に実現された）

☐ Specifically, tapped delay lines can be realized using this technology.
（具体的にいうと，タップ付き遅延線はこの技術を用いて実現できる）

☐ When the effects of delay spread are ignored, the spectral efficiency of the high-level modulation is realized only at high SNR.
（遅延拡散の影響が無視される場合，高レベル変調のスペクトル効率は高ＳＮＲでのみ実現する）

☐ There are two basic methods to realize a relaxation oscillator.
（弛張発振器を実現するための二つの基本的な方法がある）

ポイント☞［類］implement　　n:realization 実現

receive
(received, received, receiving)
〈他〉（信号などを）受信する
（注目・好意などを）受ける

☐ The radiated wave shown in the oscillogram of Fig.1 was received by the closed-loop sensor at a distane of 13 m.
（図１のオシログラムに示された放射波は，13 mの距離にある閉ループセンサによって受信される）

☐ Coherent optical transmission has <u>received</u> much attention because of higher sensitivity and many other potential advantages over intensity modulation/ direct detection (IN/DD) systems.
(コヒーレント光伝送は，輝度変調／直接検波（IN/DD)システムよりも高感度で，別の多くの潜在的な長所があるため，非常に注目されてきた)

☐ Assume that a receiving system <u>receives</u> signals from a single user via different paths.
(受信システムはさまざまな経路を経由して,一人のユーザーからの信号を受信するものと仮定する)

☐ The technology of multimedia is <u>receiving</u> lots of attention.
(マルチメディア技術は非常に注目されている)

☐ <u>Received</u> signals were amplified with a broad-band receiver having a frequency range of 0.5 to 25.0 MHz.
(受信信号は周波数範囲が0.5～25.0 MHzの広帯域受信機によって増幅された)

ポイント☞［反］transmit(送信する)　　n:reception 受信
☞receiver は受信機．heterodyne reception(ヘテロダイン受信)

record

(recorded, recorded, recording)

〈他〉（物・事を）記録する

☐ The elevation angle Θ is <u>recorded</u> at the time of each measurement.
(仰角Θは各測定時に記録される)

☐ The rf voltage and the discharge current are <u>recorded</u> using a digital storage oscilloscope and transferred to a microcomputer.
(高周波電圧と放電電流は,ディジタル蓄積型オシロスコープを用いて記録され，マイクロコンピュータへ転送される)

☐ Temperatures of the stationary components in the disk drive were measured with thermocouples whose transient signals were recorded with an automatic data acqusition system.
（ディスク駆動機構の固定部品の温度は，その過渡信号が自動データ収集システムによって記録される熱電対を用いて測定された）

☐ The analyzer can records the harmonics hourly.
（このアナライザは1時間ごとに高調波を記録できる）

ポイント☞［類］register
☞名詞としては「記録・レコード」. recording は録音・録画・記録. recorder はレコーダ・記録装置． magnetic recording(磁気記録), optical recording(光記録)， magneto-opticrecording(光磁気記録)

reduce

(reduced, reduced, reducing)

〈他〉（数量・程度・範囲などを）減らす，減少させる，低減させる
（ある物・状態などに）変える
reduce A to B：AをBに変える，変形する

☐ The effect of the random noise is reduced using the following procedure.
（ランダムノイズの影響を次の手順を用いて低減する）

☐ The current can be reduced by increasing resistance R_1.
（その電流は抵抗R_1を大きくすることで減少できる）

☐ The settling time is reduced by a factor of 10.
（整定時間は1/10に短縮した）

☐ To increase the switching frequency, the reset period of the transformer core flux must be reduced.
（スイッチング周波数を上げるためには，変圧器の鉄心における磁束のリセット周期を短くしなければならない）

☐ However, as the stability of convergence strongly depends on the circuit structure, these methods cannot reduce CPU time enough for large scale circuit simulation.
(しかしながら，収束の安定性は回路構成に強く依存するので,これらの方法では大規模回路シミュレーションに足りるだけのＣＰＵ時間を短縮できない)

☐ After A/D conversion, low-pass filtering is used to reduce the influence of noise.
(Ａ／Ｄ変換後に,低域フィルタリングが雑音の影響を低減するために用いられる)

☐ The dynamic equivalent circuit can be reduced to the steady state circuit shown in Fig.2.
(この動的等価回路は，図２に示してある定常状態回路に変えることができる)

ポイント☞ ［反］increase(増加させる)　　［類］decrease,diminish,lessen
n:reduction 減少・縮小

reflect
(reflected, reflected, reflecting)
〈他〉（光・音・熱などを）反射する

☐ These waves are reflected from the end of the piezoelectric plate.
(これらの波は圧電板の端で反射される)

☐ The beam is reflected parallel by a beam-splitting prism onto a PSD.
(ビームはビーム分割用プリズムによってＰＳＤ上に平行に反射する)

☐ As a result,both components of the incident beam are reflected with a polarization rotation of 90°.
(その結果，入射ビームの両成分は90°の偏光回転で反射する)

☐ One of the filters reflects the blue colour component of the projection lamp, while the other reflects the red one.
(フィルタの一つは映写用電球の青色成分を反射し，もう一つの方は赤色成分を反射する)

☐ The interaction of the incident wave with the plane containing the cracks gives rise to a complicated pattern of reflected and transmitted waves.
(入射波と亀裂のある面との相互作用は，反射波と透過波の複雑なパターンを引き起こす)

ポイント☞n:reflection 反射　　adj:reflective 反射する
☞reflector は反射器・反射鏡，reflectometer は反射率計．reflection coefficient(反射係数)，reflection factor(反射率)

relate

(related, related, relating)

〈他〉（物・事を）関係づける
relate A to (with) B：AをBと関係づける
〈自〉（物・事が）関係がある
relate to A：Aに関係がある

☐ The noise is related to the structure of the films.
(その雑音は膜の構造に関係する)

☐ Equation (1) relates the voltage to the average rate of change of phase.
(方程式(1)は，電圧を位相の平均変化率と関係づける)

☐ These programs aim to enhance the students' understanding of power system topics and relate them to practical problems.
(これらのプログラムは電力系統に関する論題についての学生の理解を向上させ，その論題を実際の問題に関係づけることを目指している)

☐ Eqns.2, 3 and 4 relate to narrow-beam antenna radar.

(方程式2，3，4は，狭ビームアンテナレーダに関係している)

ポイント☞n:relation 関係,relationship 関係　adj:relative 関係のある

remain

(remained, remained, remaining)

〈自〉（物・事が)とどまる，～のままでいる

☐ The motor's magnetic field remains constant during operation.

(その電動機の磁場は運転中は一定のままである)

☐ The radar cross section of reflectors remains essentially unchanged, but the concept of radar cross section cannot be applied in the usual form to scattered waves.

(反射体のレーダ断面積は本質的には変化しないが，レーダ断面積の概念は散乱波には通常の形では適用できない)

☐ Therefore, if I_n/I_c is to remain constant, I_c must increase in proportional to the operating temperature.

(したがって，I_n/I_cを一定のままにしておこうとするならば，I_cは動作温度に比例して増加しなければならない)

☐ The Faraday rotation remained nearly constant over the considered temperature range.

(ファラデー回転は考慮に入れている温度範囲にわたって，ほぼ一定のままだった)

ポイント☞［類］stay

☞名詞としては「残り・残留品(通例複数形)」.

remove

(removed, removed, removing)

〈他〉(不要なものを)取り除く，除去する

remove A from B：AをBから取り除く

☐ This restriction is readily removed in a fully monolithic design.
(この制限は完全モノリシック設計では容易に取り除ける)

☐ The residual photoresist is removed using an oxygen plasma.
(残ったフォトレジストは酸素プラズマを用いて取り除かれる)

☐ These errors are known as systematic errors which can be removed by a calibration procedure.
(これらの誤差は,校正手順によって取り去ることができる系統的誤差として知られている)

☐ Fluid is removed from the tank at a constant 1 m^3mim^{-1}(±10%) from the outlet pipe.
(流体は出口パイプから1 m^3mim^{-1}(±10%)の割合でタンクから出される)

☐ The procedure removes the dark noise of the image intensifiers.
(その手順でイメージインテンシファイヤーの暗雑音を除去する)

☐ The filter can effectively remove background noise from smooth areas of the signal.
(そのフィルタは,信号の平滑領域からバックグラウンドノイズを効果的に除去できる)

☐ The purpose was to design a low-pass filter in order to remove the high-frequency noise.
(この目的は,高周波雑音を除去するために低域フィルタを設計することである)

ポイント☞［類］eliminate,get rid of　　n:removal 除去
☞「必要ない物を動かして取り除く」のニュアンスがある.

repeat

(repeated, repeated, repeating)

〈他〉（動作・行為などを）繰り返す

□ This procedure is repeated periodically.
(この手順は周期的に繰り返される)

□ The radiation properties of the TM_{21} mode for the proposed element are similar to those of the conventional disk antenna,and are not repeated in this paper.
(提案された素子に対するTM_{21}モードの放射特性は，従来のディスクアンテナの放射特性に似ているので，本論文では繰り返さない)

□ Similar tests have been repeated with a variety of circuits.
(同様の試験がさまざまな回路を用いて繰り返されてきた)

□ The calibration procedure described in this paper should be repeated for a series of different wavelengths in the wavelength range of interest.
(本論文で記述した校正手順は,考察の対象となっている波長領域における一連の異なった波長に対して繰り返すべきである)

□ We repeat the procedure from equ.(8) as described above until the resulting Josephson current density distribution converges to a final value.
(その結果得られたジョセフソン電流密度分布が最終値に収束するまで,上記のごとく方程式(8)の手順を繰り返す)

□ Two dimensional space resolution is obtained by repeating the measurements with the chamber moved in the x direction.
(チャンバをx方向に移動させ,測定を繰り返すことで二次元空間分解能を求める)

ポイント☞n:repetition 繰り返し　　adj:repeated 繰り返された
adv:repeatedly 繰り返して
☞名詞としては「繰り返し・反復」. repeater は中継器. request repeat system(再送訂正方式)

replace

(replaced, replaced, replacing)

〈他〉（物・事に）取って代わる
replace A with (by) B：AをBと取り替える，交換する

□ The future outputs are replaced by the optimal predictions.
（未来の出力値を最適予測値と取り替えた）

□ Nyquist noise in the shunt resistors must be replaced with zero point fluctuations.
（分路抵抗器のナイキストノイズは，零点ゆらぎと取り替えなければならない）

□ Non-contact laser transducers can replace contact piezoelectric probes for many applications.
（多くの用途にとって，非接触レーザ変換器は接触型圧電プローブに取って代われる）

□ In many applications, optical methods are replacing mechanical and electro-mechanical methods.
（多くの用途において，光学的方法は機械的および電気機械的方法に取って代わりつつある）

ポイント ［類］substitute,supplant　　n:replacement 交換

report
(reported, reported, reporting)
〈他〉（研究・調査などを）報告する

□ A number of approaches have been reported to simplify the implementation of one-dimensional FIR filters.
（一次元ＦＩＲフィルタの実現を容易にするために，多くのアプローチが報告されている）

□ A significant amount of research has been reported on adaptive nonlinear filters.
（適応非線形フィルタについて，著しい数の研究が報告されている）

□ We report the first low-noise InP/InGaAs heterostructure bipolar transistor (HBT).
(最初の低雑音InP/InGaAsヘテロ構造バイポーラトランジスタ(HBT)を報告する)

□ Several researchers have reported methods for the rapid assessment of reliability of multilayer ceramic capacitors under varying conditions of accelerating voltage.
(加速電圧の変化する条件のもとで,多層磁器コンデンサの信頼性をすばやく評価するための方法を，幾人かの研究者が報告している)

□ We show some of the capabilities of the derived algorithms by reporting their application in simulation to a multirate sampled data system.
(導出したアルゴリズムの能力のいくつかを,マルチレート・サンプル値システムに対するシミュレーションでの応用を報告することによって示す)

ポイント☞adv:reportedly 伝えられるところによると
☞名詞としては「報告(書)」. 名詞の他に動名詞・that節も目的語にできるが，不定詞は不可.

represent

(represented, represented, representing)

〈他〉（物・事を）表す，表現する

□ According to (2),the character of a binary code is represented by a Gaussian pulse of duration ΔT and peak-amplitude E.
((2)によると，２進コードの文字は持続時間ΔT, ピーク振幅Ｅのガウスパルスで表現される)

□ Active elements such as FETs are usually represented by their equivalent circuit models.
(ＦＥＴのような能動素子は，通常それらの等価回路モデルで表現される)

□ Time-varying signals may be represented efficiently using orthogonal

transformations.
（時変信号は直交変換を用いて効率よく表現できる）

□ The Walsh transform can be represented in a matrix form.
（ウォルシュ変換は行列形式で表現できる）

□ The capacitances C_p represent the parasitic capacitances of the windings of the choke.
（静電容量C_pはチョークの巻線の寄生容量を表す）

ポイント☞［類］express, indicatedenote　n:representation 表現　adj:representative 表現する・代表する
☞digital representation(ディジタル表現), knowledge representation(知識表現)

require
(required, required, requiring)
〈他〉（物・事を）必要とする，要求する
require that節：～ということを必要とする，要求する

□ Reduction of the linewidth of conventional long wavelength laser diodes is required for thier use in coherent communications systems.
（従来の長波長レーザダイオードの線幅の縮小は，コヒーレント通信システムでの使用に必要とされる）

□ The heater control circuit in Fig.2 is required for operation and keeps the heater temperature at a constant above ambient temperature.
（運転には図２のヒータ制御回路が必要で，この回路はヒータの温度を周囲温度以上のある一定の値に維持する）

□ A 256 kbit DRAM requires 18 address bits to address each data bit.
（256 Ｋビット DRAMは，各データビットをアドレス指定するためには18アドレスビットが必要である）

□ The numerical solution of integral equations by the method of moments usually requires the evaluation of Green's functions which involve computation of integrals.
(モーメント法による積分方程式の数値解法には,普通は積分の計算を含むグリーン関数の数値計算が必要である)

□ In comparison to other methods, the tanh transformation does not require the use of any special transforms, such as FFT or CZT.
(他の方法と比較して,この双曲線正接変換にはFETあるいはCZTのような特別な変換を使用する必要はない)

□ High Tc SQUID systems do not require liquid helium cooling systems.
(高Tc ＳＱＵＩＤシステムには液体ヘリウム冷却システムは必要ない)

□ The basic principle of metrology requires that the reference standard instrument must have an accuracy of one step better than that one of the tested instrument.
(計測学の基本的原理は,照合標準器の精度は試験される計器の精度よりも一段階上でなければならないということが要求される)

ポイント☞［類］need　　n:requirement 必要(条件)
☞進行形は不可. software requirement specification(ソフトウェア要求仕様)

restrict

(restricted, restricted, restricting)

〈他〉（事を）制限する，限定する

□ The computation will be restricted to the conductor between the two electrodes.
(この計算は，２個の電極間の導体に限られる)

□ To illustrate the procedure, we will restrict ourselves to a specific example that has proven to be stable in all cases.
(この手順を例示するために,あらゆる場合に安定であることが証明されているある特定の例に限定する)

☐ It is important to restrict dislocations to the interfacial region between the two materials.
(二つの材料間の界面領域に転位を制限することは不可能である)

ポイント☞［類］limit, confine　　n:restriction 制限・限定　　adj:restrictive 制限する
☞「ある範囲内に限定し，そこから出ないようにする」のニュアンスがある．

result
(resulted, resulted, resulting)
〈自〉（物・事が）生じる
result from A：Aから生じる，Aに起因する
result in A：Aの結果になる，Aをもたらす

☐ If current is distributed to the junction above the voided area, a hot spot can result.
(もし電流が間隙領域上の接合部に分布すれば，ホットスポットが生じる)

☐ The linearity of the input stage results from the measured harmonic distortion.
(入力段の線形性は測定された高調波歪みに起因する)

☐ Image speckle results from the interference patterns generated by random scatterings from rough surfaces.
(画像のスペックルは粗面からの不規則散乱によって発生した干渉パターンが原因で生じる)

☐ The use of the hybrid coupler results in high stability.
(ハイブリッド結合器を使用すると，高安定性が得られる)

☐ At sufficient power levels, this nonuniform temperature profile results in hot spot formation which can leads to thermal runaway.
(十分な電力レベルでは，この不均一な温度プロファイルのために熱暴走を引

き起こしかねないホットスポットが形成される）

□ The drop in current with increasing positive bias results in the negative characteristic of the tunnel diode.
（正バイアスの増加に伴う電流の降下は，トンネルダイオードの負特性をもたらす）

□ One of the fransformers can be over-saturated, resulting in higher core loss.
（変圧器の一つが過飽和する場合があり，その結果，大きな鉄損が生じる）

□ Therefore,the turn-on switching loss is eliminated, resulting in high efficiency of DC/DC energy conversion at high frequencies.
（したがって，ターンオンスイッチング損失が除去されると，高周波における DC/DCエネルギー変換の効率が高くなる）

ポイント☞［類］arise,happen,occur,take place,give rise to　　adj:resultant 結果としての
☞名詞としては「結果」．A results from B.＝B results in A.（AがBから生じる＝BはAの結果になる）の関係に注意．The resultant (resulting)～は「その結果生じた～」．

reveal
(revealed, revealed, revealing)
〈他〉（ある事を）明らかにする，示す
reveal that節：～であることを明らかにする，示す

□ The high resolution micrograph reveals a cross section of the interface having a roughness of less than 3 nm.
（高解像度マイクログラフによって，粗さが3nm以下の界面の断面を示す）

□ It is the purpose of this paper to reveal the essential properties of these two topologies and to compare them.
（これら二つのトポロジーの本質的な特性を明らかにし，両者を比較することが本論文の目的である）

☐ Investigation revealed that the noise is due to the frequency instability of the VCO.
(その雑音がVCOの周波数不安定性によって生じるということを，研究によって明らかにした)

☐ Examination of (19) reveals that the sensitivity is a function of frequency and permittivity.
((19)を検討することで,この感度が周波数と誘電率の関数となることを明らかにする)

ポイント☞［類］clarify, disclose, elucidate, exhibit, show, display, demonstrate, indicate, illustrate
☞「今まで知られていなかったことを明らかにする」のニュアンスがある.

run

(ran, run, running)

〈他〉 (プログラムを)実行する
(機械・装置などを)動かす
〈自〉 (プログラム・機械・装置などが)動作する
(ケーブル・配線などが)通る

☐ The SCAN program is written in C programming language and run on IBM personal computers.
(SCANプログラムはCプログラミング言語で書かれており,IBMパーソナルコンピュータで実行される)

☐ The preprocessor was run using the edited file as input.
(そのプリプロセッサは編集されたファイルを入力として用いることで稼働する)

☐ This has impacted the industrial workstations on which those software packages run.

(このことは,それらのソフトウェアパッケージが動作する工業用ワークステーションに影響を与えた)

☐ Signal cables ran from the interface to each of the gyroscopes and to the IBM PC.
(信号ケーブルはインタフェースから各々のジャイロスコープまで,およびＩＢＭ　ＰＣまで引かれた)

☐ The rotary kiln in a cement industry is required to run on a speed range of 1/3 to 1.8 rpm.
(セメント工業におけるロータリー窯は，1/3～1.8 rpmの速度範囲で動作することが必要である)

ポイント☞［類］execute, move, work,do, perform, make
☞名詞としては「実行・走行」.

S

sample

(sampled, sampled, sampling)

〈他〉（信号などを)標本化する，サンプリングする

☐ The multiplier output is sampled by an amplitude detector whose output is lowpass-filtered.
(乗算器の出力は,その出力が低域ろ波される振幅検出器によってサンプリングされる)

☐ The waveform of voltage or current was sampled and digitized, and a fast Fourier transform (FFT) was performed to determine the harmonics.
(電圧，あるいは電流の波形はサンプリング，およびディジタル化され,高調

波を決定するために高速フーリエ変換(ＦＦＴ)を行った)

□ The electrical signal supplied by the interferometer corresponding to the ultrasonic displacement is sampled by a digital oscilloscope.
(超音波変位に対応する干渉計によって供給された電気信号は，ディジタルオシロスコープによって標本化される)

□ These systems must sample signals at higher frequencies.
(これらのシステムは高周波で信号を標本化しなければならない)

ポイント☞n:sampling 標本化・標本抽出・サンプリング
☞名詞としては「標本・サンプル」. sampled data control system(サンプル値制御系), sampling theorem(標本化定理)

satisfy
(satisfied, satisfied, satisfying)
〈他〉（要件・条件などを）満たす，満足させる

□ S_1 is reachable if and only if the following two conditions are satisfied.
(次の二つの条件が満たされる時，かつその時に限ってS_1は可到達となる)

□ If this basic condition is not satisfied, then this characterization of the optimum solution is no longer valid.
(この基本条件が満たされないならば，最適解のこの特性づけはもはや有効ではない)

□ It is known that the orthogonal filter structures satisfy $K=I_n$.
(その直交フィルタ構造が$K=I_n$を満足させるということは知られている)

□ These processes, known as semi-Markov chains, do not necessarily satisfy the strict Markovian property.
(半マルコフ連鎖として知られているこれらの過程は，必ずしも狭義のマルコフ性を満たさない)

☐ SIS Junctions for mm-wave mixers should satisfy the following conditions to realize the ultra-low noise property.
（ミリ波ミクサー用のＳＩＳ接合は，超低雑音特性を実現するために次の条件を満たさなければならない）

☐ This paper proposes a new approach which produces module generators satisfying all the requrements.
（本論文では，すべての要件を満たしているモジュールジェネレータを生成する新しいアプローチを提案する）

ポイント☞［類］meet, fulfill　　n:satisfaction 満足　　adj.satisfactory 満足な
☞通例進行形は不可．

scatter

(scattered, scattered, scattering)

〈他〉（電磁波・光などを）散乱させる

☐ The electromagnetic field was scattered by a perfect conductor.
（電磁場は，完全導体によって散乱された）

☐ That is, we cannot calculate the backscatter coefficient which describes how the grains scatter the sound.
（すなわち，この粒子がどのように音を散乱させるかということを記述する後方散乱係数を計算することはできない）

☐ The electromagnetic fields scattered outside and inside the dielectric cylinder can readily be determined in terms of the equivalent currents.
（誘電体円柱の外部と内部で散乱した電磁場は，等価電流により容易に決定できる）

☐ A coupled finite boundary element method is developed for the EM fields scattered by an arbitrary multiple-dielectric coated conducting cylinder.
（任意の多層誘電被覆導体円柱によって散乱した電磁場のために，結合有限要素法を開発する）

ポイント☞名詞としては「散乱」. scattering も散乱. scatterer は散乱体. scattering coefficient(散乱係数), scattering cross section(散乱断面積), scattered light(散乱光)

see

(saw, seen, seeing)

〈他〉（事が）わかる

see that節：～ということがわかる

□ Agreement between theory and experiment is <u>seen</u> to be quite good.
（理論と実験との間にかなりよい一致があることがわかる）

□ As can be <u>seen</u> from Fig.5, the agreement of the analysis and the experiment is not good, even though the resonance frequencies have been predicted well.
（図5からわかるように，たとえ共振周波数が十分に予測されていても，解析と実験の一致はよくない）

□ It is <u>seen</u> that problems arise if the material is very thin.
（その材料が非常に薄ければ問題が生じるということがわかる）

□ When measuring materials with a low dielectric constant, it is <u>seen</u> in Fig.6 that the influence of higher order modes is very important if the material is thin.
（誘電率が低い材料を測定する場合，その材料が薄ければ高次モードの影響は非常に重要となるということが，図6からわかる）

ポイント☞［類］prove, turn out

☞「洞察や推論の力でもってわかる」のニュアンスがある.

select

(selected, selected, selecting)

〈他〉（物・事を）選択する

select A to do：Aが～するように選ぶ

select A as B：AをBとして選ぶ

☐ Low drift operational amplifiers are selected for the measuring circuit.
（低ドリフト演算増幅器をこの測定回路用に選ぶ）

☐ The concentration of the reactive gases was selected using electronic mass flow controllers.
（その反応性ガス濃度を，電子式質量流量制御器を用いて選択した）

☐ The reasons we selected an HBT to provide gain are the following.
（利得を与えるためにＨＢＴを選んだ理由は次の通りである）

☐ An analog switch selects the appropriate modulation frequency to mix with the carrier (44 kHz).
（アナログスイッチは，適切な変調周波数(44 kHz)を搬送波と混合するように選ぶ）

☐ We selected SiO_2 as an isotropic material.
（SiO_2を等方性物質に選んだ）

☐ Two methods to select the ADP density pattern from possible patterns were investigated.
（可能なパターンからＡＤＰ密度パターンを選択する二種類の方法について調べた）

☐ Before selecting a pressure gauge for its accuracy, one need to understand what accuracy means in these devices.
（その精度に対して圧力計を選択する前に，これらの装置において精度が何を意味するかを理解する必要がある）

ポイント☞［類］choose, prefer　n:selection 選択　adj:selective 選択的な　adv:selectively 選択的に
☞「その目的を十分に考え，慎重な判断で選択する」のニュアンスがある．

send

(sent, sent, sending)

〈他〉（信号・情報などを）送る，送信する

□ The EGM signal from each electrode is sent through the differential amplifier stage.
（各電極からのＥＧＭ信号は差動増幅器段を介して送られる）

□ The computer can send the results to a display or to a control system for further processing.
（そのコンピュータは結果を表示装置，またはさらに処理するために制御システムへ送ることができる）

□ The D/A converter converts the digital signal from the IIR filter into the analog form and sends it to TV monitor.
（そのＤ／Ａ変換器はIIRフィルタからのディジタル信号をアナログ方式に変換し，その信号をTVモニターに送る）

□ The resulting signal is processed by the low-pass filter and sent to the comparator.
（その結果得られた信号を低域フィルタで処理し，コンパレータへ送る）

□ The phase difference between two square waves is obtained by sending them through an exclusive OR gate.
（２個の方形波間の位相差は，その方形波を排他的論理和ゲートを介して送ることで得られる）

ポイント☞［反］receive(受信する)　　［類］transmit
☞「ある場所から別の場所へ移動させる」のニュアンスがある．

sense

(sensed, sensed, sensing)

〈他〉（状態・変化などを）検知する，感知する

☐ This current is sensed by the current mirrors.
（この電流はカレントミラーによって検知される）

☐ This technology is useful when trying to sense phenomena such as vibration and fluid flow.
（この技術は，振動や流体流といった現象の検知を試みる時に有用である）

☐ This effect can very well be applied to sense high hydrostatic pressure using appropriately configured HB fiber, but it could as well be a disadvantage in underwater communication systems employing HB fibers if these are not adequately protected.
（この効果は，適切に構成されたＨＢファイバを用いて高静水圧を検知するのに大変うまく応用できるが，ＨＢファイバを用いた水中通信システムで，もしこのファイバが十分に保護されなければ，不利点もまた有する可能性もある）

☐ A strain gauge will be used to sense cable tension,and an encoder will sense payload position.
（歪みゲージは，ケーブルの張力を検知するのに用いることができ，エンコーダはペイロードの位置を検知するのに用いることができる）

☐ The optical fiber interferometer was used to sense Rayleigh waves in the presence of significant low-frequency ambient disturbances due to temperature variations, noise, and vibrations.
（光ファイバ干渉計は，温度の変動，雑音，振動による大きな低周波周囲擾乱の存在のもとで，レーリー波を検知するのに用いられた）

ポイント☞adj:sensitive 感度のよい
☞名詞としては「感覚・意味」. 通例進行形は不可. sensor はセンサ. in a sense は「ある意味では」, make sense of は「～を理解する」. temperature sensor(温度センサ), image sensor(イメージセンサ), smart sensor(スマートセンサ)

separate

(separated, separated, separating)

〈他〉（物を）分離する，分ける

separate A from B：AをBから分ける

□ The heater is completely separated from the laser.

（ヒータはレーザから完全に隔てられている）

□ The output can be separated from the input with a beamsplitter.

（その出力はビームスプリッタによって入力から分離できる）

□ In this method,the thin wires are separated from the surrounding mesh through the use of MoM.

（この方法の場合，MoMの使用によって細線はその周囲のメッシュから分けられる）

□ The Weiss domains are separated by Bloch walls,in which the magnetization gradually changes from the direction in one domain to that in the neighbouring domain.

（ワイス領域はブロッホ壁で分離され，そこでは磁化は一つの磁区方向から隣接した磁区方向へ徐々に変わる）

□ The real coil has a finite lenght and its windings are separated by air gaps.

（実際のコイルは有限の長さをもち，その巻線はエアギャップで隔てられている）

□ An airgap of thickness *s* separates the two dielectric layers.

（厚さが*s*のエアギャップは，二つの誘電層を分離している）

ポイント☞［類］isolate　n:separation 分離　adv:separately 別々に

☞「一体であったものを引き離して別々の物にする」のニュアンスがある．

set

(set, set, setting)

〈他〉（物・事を）設定する，セットする

set up A：Aを組み立てる，立てる，もたらす

□ The working temperature can be set by means of a trimmer in module B.
（使用温度はモジュールBのトリマによって設定できる）

□ The polarization controller can be set so that the output power of the attenuator becomes a desired value.
（分極制御器は，減衰器の出力電力が所望の値となるように設定できる）

□ The resistance R_a was set at 6.8 MΩ for minimum noise and maximum signal output.
（抵抗R_aは雑音が最小で信号出力が最大となるように，6.8 MΩに定められた）

□ The •detector surface is initially set normal to the laser beam by autocollimation, so that the reflected beam coincides with the incident beam at $\phi=0$.
（検出器の表面はオートコリメーションによってレーザビームに対して垂直となるように初期設定され，そのため$\phi=0$において反射ビームは入射ビームと一致する）

□ Suppression level may be set by varying the ratio of R_1 to R_2.
（抑制レベルはR_1とR_2の比を変えることでセットできる）

□ The test system was set up as shown in block diagram in Fig.1.
（試験システムは図１のブロック線図に示すように組み立てた）

□ If a plasma is present, a space charge will be set up, and the potential will not satisfy Laplace's equation.
（プラズマが存在するならば，空間電荷が生じ，電位はラプラス方程式を満たさなくなる）

ポイント☞名詞としては「一組・セット・集合」. setting は設定.
☞fuzzy set(ファジイ集合)

shift
(shifted, shifted, shifting)
〈他〉(物・事を)移す，移動させる
(ビット列・文字列などを)桁送りする

☐ The sample surface is twice shifted by exactly λ/8 and the three resulting interference patterns are recorded.
(試料面はちょうどλ/8だけ2回移動され，得られた三つの干渉縞が記録される)

☐ Computer simulation predicted that the shift register could shift data at 12 ps/bits if the Josephson critical current density were $j_c = 10$ kA/cm^2.
(もしジョセフソン臨界電流密度が$j_c = 10$ kA/cm^{2}ならば，シフトレジスタはデータを12ps/ビットでけた送りできるということを，コンピュータシミュレーションで予測した)

☐ The basic approach to the solution of the estimation problem is to shift X(n) with respect to Y(n).
(その推定問題を解くための基本的アプローチは，Y(n)に対してX(n)を移すことにある)

ポイント☞［類］move, migrate, transfer, transport
☞名詞としては「移動・シフト」. shift register(シフトレジスタ)

show
(showed, showed, showing)
〈他〉(物・事を)示す，表す
show that節：～ということを明らかにする，証明する，示す

☐ The equivalent transfer function is shown in Fig.3.
(等価伝達関数を図3に示す)

☐ Figure 1 shows typical examples of speech waveforms.
(図１に，音声波形の代表例を示す)

☐ Many weak couplings of two superconductors show the Josephson effect.
(２個の超伝導体の弱結合の多くは，ジョセフソン効果を示す)

☐ This panel shows no evidence of either phosphor degradation or panel voltage instability.
(このパネルは蛍光体の劣化，あるいはパネル電圧の不安定性のどちらの形跡も示さない)

☐ Computer simulation shows that the performance of the proposed algorithm is better than existing ones in terms of bit rate and subjective image quality.
(コンピュータシミュレーションにより，提案されたアルゴリズムの性能はビットレートと主観的画質の点で従来のよりも優れているということを明らかにする)

☐ The goal of this paper is to show that the following theorem 3 is true.
(本論文の目的は，次の定理３は真であるということを証明することである)

ポイント☞［類］exhibit, display, reveal, demonstrate, indicate, illustrate, present
☞名詞としては「展示会」.

simplify
(simplified, simplified, simplifying)
〈他〉（物・事を）単純化する，簡単にする，簡略化する

☐ Equations (2) and (3) are simplified using Laplace transform techniques.
(方程式(2)と(3)を，ラプラス変換法を用いて単純化する)

☐ The mathematical description of the magnetic field was considerably simplified.
(磁場の数学的記述はかなり単純化された)

☐ The user interface with the harmonic analyzer is greatly simplified by the use of

a personal computer.
(調波分析器とのユーザーインタフェースは,パーソナルコンピュータの使用によって大いに簡素化される)

☐ This simplifies the measurement technique and eliminates several sources of error.
(これによって測定手法は簡単になり，いくつかの誤差の原因は取り除かれる)

ポイント☞n:simplification 単純化　adj:simple 単純な　adv:simply 簡単に

simulate
(simulated, simulated, simulating)
〈他〉（特性・機能・挙動などを)シミュレートする

☐ Transient step response was simulated using the TEM mode lossless transmission line algorithm discussed earlier.
(過渡ステップ応答は,先に論じられたＴＥＭモード無損失伝送線路アルゴリズムを用いてシミュレートされた)

☐ The robust control was simulated using SIMNON, and the simulation results are shown in the figures.
(ロバスト制御をSIMNONを用いてシミュレートし,そのシミュレーション結果は図に示している)

☐ In order to study the effect of the operational amplifier parameters on the oscillator circuit,we have simulated the oscillator using the SPICE program.
(演算増幅器のパラメータが発振器回路に及ぼす影響を研究するために,SPICEプログラムを使用して発振器をシミュレートした)

☐ To simulate the accelerometer performance by computer, a mathematical model that considers the electrical and mechanical properties of the device must be developed.
(加速度計の性能をコンピュータでシミュレートするために,その装置の電気

的および機械的性質を考慮に入れた数学モデルを開発しなければならない）

□ In the section we present some simulated examples to evaluate the performance of the presented adaptive control scheme.
（この節では，提出された適応制御方式の性能を評価するために模擬例を呈示する）

ポイント☞n:simulation シミュレーション
☞simulator はシミュレータ． simulation model(シミュレーションモデル)， computer simulation(コンピュータシミュレーション)， simulator program(シミュレータプログラム)， flight simulator(フライトシミュレータ)

solve
(solved, solved, solving)
〈他〉（問題・数式などを）解く

□ These problems can be solved in a number of ways.
（これらの問題はいろいろな方法で解くことができる）

□ This integral equation was solved numerically by the method of moments.
（この積分方程式は，モーメント法によって数値的に解かれた）

□ It has been shown that Hopfield-type neural networks can solve a number of difficult optimization problems in a time determined by system RC time constants.
（ホップフィールド型ニューラルネットワークは，システムのＲＣ時定数によって決定される時間内で多くの難解な最適化問題を解くことができるということが明らかになっている）

□ The Finite-Difference Time-Domain (FDTD) algorithm is used to solve these equations in the time domain.
（有限差分時間領域（FDTD）アルゴリズムは，時間領域でこれらの方程式を解くのに用いられる）

☐ One way of solving this problem is the use of modulation techniques having high spectral efficiency.
(この問題を解くための一つの方法は,高スペクトル効率を有する変調手法を使用することである)

☐ The method of moments is used for solving both the partial differential equations for the interior region and the integral equations for the exterior region.
(モーメント法は，内部領域に対する偏微分方程式,外部領域に対する積分方程式の両方を解くために使用される)

☐ In recent years, a large variety of methods has been proposed for solving the failure detection problem.
(近年，故障検出問題を解決するために,実にさまざまな方法が提案されてきた)

☐ Solving the boundary conditions at the dielectric-ferrite interface the normalized impedance matrix elements of the DF circulator are derived.
(誘電体・フェライト界面での境界条件を解けば,ＤＦサーキュレータの正規化インピーダンス行列の要素が導出される)

ポイント☞n:solution 解決
☞approximate solution(近似解), periodic solution(周期解)

stabilize

(stabilized, stabilized, stabilizing)

〈他〉（物・事を)安定化する，安定させる

☐ The interferometer was stabilized against room vibrations.
(干渉計を部屋の振動に対して安定化させた)

☐ We must design a robust controller to stabilize the closed-loop system.
(閉ループシステムを安定化させるために,ロバスト制御器を設計しなければならない)

☐ W and R are feedback signals, which are needed to stabilize the adaptive observer.
(WとRはフィードバック信号であり，これらは適応オブザーバを安定化するのに必要である)

☐ In order to stabilize the DFB-LD,backward reflected light must be eliminated.
(DFB-LDを安定化するには，後方反射光を除去しなければならない)

☐ The vibrational control has been used for stabilizing chemical reactor operation.
(化学反応器の運転を安定化するために，振動制御が用いられた)

ポイント☞n:stability 安定性，stabilization 安定化　adj.stable 安定した
☞stabilizer は安定装置・安定剤．frequency stability(周波数安定度)，stability analysis(安定性解析)

start

(started, started, starting)

〈他〉 (事を)始める
start doing/to do:～し始める
〈自〉 (事が)始まる
start with A:Aで始まる

☐ We start this section with a brief overview of notation and some terms used in the text.
(本節では，本文で使用される表記法と用語の簡単な概略から始める)

☐ The station starts transmitting messages.
(ステーションはメッセージを伝送し始める)

☐ The film starts becoming optically transparent in the infrared (＞800 nm).
(膜は赤外線(＞800 nm)領域で光学的に透明になり始める)

☐ We start to encode the sequence of information bits from time $k=1$ to $k=N$.

（$k=1$から$k=N$まで，情報ビットのシーケンスの符号化を開始する）

□ The course starts with an overview of communication networks.
（このコースは通信網の概要で始まる）

□ We start by giving a brief introduction to concurrency control.
（まず始めに，平行処理制御について簡単に紹介する）

ポイント☞［類］begin,initiate
☞「新たにある動作や行為を始める」のニュアンスがある．主語が物の場合は，start to do のほうが一般的である．名詞としては「開始・出発」．to start with(まず第一に，初めは)

state
(stated, stated, stating)
〈他〉（事実・意見・問題などについて）述べる，示す
state that節：～ということを述べる，明言する

□ The robust control problem is stated in Section Ⅲ.
（第３節では，ロバスト制御問題について述べられる）

□ The control problem we are dealing with can be stated as follows.
（扱っている制御問題は，次のように述べることができる）

□ We state some interesting conjectures.
（いくつかの興味深い推測について述べる）

□ The DSCS states that the backscatter from $N+1$ cracks is equal to the sum of the backscatter from N cracks, plus the backscatter from a single crack in an effective interphase.
（$N+1$個の亀裂からの後方散乱は，N個の亀裂からの後方散乱に有効中間相の１個の亀裂からの後方散乱を加えたものに等しいということを，ＤＳＣＳは示している）

☐ As stated in the introduction, we consider systems described by difference equations.
(序文で述べたように，差分方程式で記述されるシステムについて考察する)

ポイント☞［類］describe　n:statement 声明・述べること
☞「意見・理由などを言葉で明確に述べる」のニュアンスがある．名詞としては「状態」．state of the art(現在の[科学]技術水準)

store
(stored, stored, storing)
〈他〉（データを）格納する，蓄積する，保管する，記憶する
（エネルギーなどを）貯蔵する

☐ Process data is stored on hard disk and displayed on a CRT monitor using either a keyboard or track-ball for selection.
(プロセスデータはハードディスクに保管され，選択用にキーボードかトラックボールのどちらかを使ってＣＲＴモニターに表示される)

☐ Single quanta of magnetic flux are stored as persistent currents in a loop containing two Josephson junctions.
(単一の磁束量子は，２個のジョセフソン接合を含むループ中に永久電流として蓄えられる)

☐ PLCs and other control devices store data in registers.
(PLCと他の制御装置はレジスタにデータを格納する)

☐ Modern computers store data in files.
(現代のコンピュータはデータをファイルに格納する)

☐ The latest optical disks can store a few gigabytes.
(最近の光ディスクは数ギガバイトの記憶能力がある)

☐ As mentioned before, the memories are used to store all image data.
(すでに述べたように，このメモリはすべての画像データを格納するのに用い

られる）

☐ In this section we derive an expression for energy stored in a lossless network.
（この節では，無損失ネットワークに蓄積されるエネルギーのための数式を導出する）

ポイント☞n:storage 貯蔵・記憶(装置)
☞名詞としては「蓄積・記憶(装置)」. disk store[storage](ディスク記憶装置), mass store[storage](大容量記憶装置), main store[storage](主記憶装置)

study
(studied, studied, studying)
〈他〉（物・事を）研究する，調べる

☐ The stability of this system can be studied by using the following Lyapunov function.
（このシステムの安定性は，次のリャプノフ関数を用いることで研究できる）

☐ During the last twenty years the regulator problem was extensively studied by several authors.
（過去２０年間に，レギュレータ問題は幾人かの著者によって広範囲にわたって研究された）

☐ In this section, we study the necessary and sufficient condition for the reachibility of S_1.
（この節では，S_1の可到達性のための必要十分条件について調べる）

☐ This model is useful to study multiuser communication systems.
（このモデルはマルチユーザー通信システムを研究するのに役立つ）

☐ Numerical techniques have been applied to study the effects of interface traps on device performance.
（界面トラップがデバイスの性能に及ぼす影響を研究するために，数値手法が

適用された）

ポイント☞［類］investigate, research, examine, explore

☞「対象を詳しく観察・分析し，綿密に調べる」のニュアンスがある．名詞としては「研究・調査」．

suffer

(suffered, suffered, suffering)

〈他〉（損害・作用などを）受ける，被る

〈自〉（物・事で）損害を被る

suffer from A：Aで損害を被る

□ The gate pulse suffers distortion as it moves along the line.

（ゲートパルスは線路に沿って移動するにつれて歪みを受ける）

□ Random noise and uncorrelated signals, which do not match the impulse response, suffer the CW insertion loss of the filter.

（そのインパルス応答に一致しないランダムノイズと無相関信号は，フィルタのＣＷ挿入損を被る）

□ This device suffered from major shortcomings such as analog I/O with limited resolution and the absence of a dedicated multiplier.

（このデバイスには，アナログＩ／Ｏには分解能に制限があり，そして専用の乗算器がないといった大きな短所がある）

□ $Hg_{1-x}Cd_xTe$ devices also suffer from 1/f noise.

（$Hg_{1-x}Cd_xTe$素子もまた1/f雑音を受ける）

□ This analog correlator is less expensive than a digital correlator and does not suffer from sensitivity degradation due to quantization of the signals.

（このアナログ相関器はディジタル相関器より安く，しかも信号量子化による感度の劣化を受けない）

ポイント☞［類］undergo　　n:suffering 苦しみ・被害

suggest

(suggested, suggested, suggesting)

〈他〉 (計画・考えなどを)示唆する，提案する
suggest that節:～ということを示唆する，提案する
suggest doing:～することを提案する

□ A new method is suggested to improve the low frequency behaviour.
(その低周波挙動を改善するための新しい方法を提案する)

□ Equations (45) and (46) suggest an easy way to construct a parameterized PLN model from the measured 1st-and 2nd-order kernels.
(方程式(45)と(46)は，測定された一次，および二次核からパラメータ化されたＰＬＮモデルを構築する容易な方法を示唆している)

□ This result suggests that (13),under certain conditions,can be used to differentiate the LN model from the LNL model.
(この結果は，ある条件下で,(13)はＬＮモデルをＬＮＬモデルと区別するのに用いることができるということを示唆している)

□ This experimental result suggests that the trap density can be controlled by optimizing the processing.
(この実験結果は,トラップ密度はその処理を最適化することで制御できるということを示唆している)

□ Some authors suggest taking temperature readings only after switching off the power generator in order to avoid the second problem.
(二番目の問題を避けるために,発電機のスイッチを切った後に温度を測定することを提案している著者もいる)

ポイント☞［類］propose, put forward　n:suggestion 示唆・提案
adj:suggestive示唆に富む
☞「控え目に提案する」のニュアンスがある．that節内の動詞は仮定法現在か should～.

suit

(suited, suited, suiting)

〈他〉（物・事に)適する，適合する

☐ The traditional cooling systems are not suited for laboratory low-noise instrumentation.
(従来の冷却システムは研究室の低雑音計装には適していない)

☐ Gaussian pulse model is well suited for the analytical study of radar signal processing.
(ガウスパルスモデルはレーダの信号処理の解析的研究にたいへん適している)

☐ This kind of analysis is suited to the treatment of long spans of fiber, where the ordinary eigenmode description becomes useless because of the coupling increase with the propagation length.
(この種の解析は長径間ファイバを取り扱うのに適しており，そしてここでは伝搬の長さとともに結合が増加するため，通常の固有モード記述は役に立たなくなる)

☐ The control method has been shown to be well suited to the problem of obstacle avoidance in robotics.
(この制御法はロボット工学における障害物回避問題にたいへん適しているということが示されている)

ポイント☞adj:suitable 適した
☞suit は多くは be suited (to,for)の形で使われる．suit＝be suitable for

summarize

(summarized, summarized, summarizing)

〈他〉（物・事を)要約する

☐ The modelling procedure is summarized at the end of the section.

（モデリングの手順はこの節の最後に要約する）

□ The theoretical model is described,numerical results and a discussion are presented, and finally the conclusions are summarised.

（理論モデルについて述べ，数値結果と考察を示し，最後に結論を要約する）

□ A simple procedure is summarized for computing the electromagnetic diffraction from a dielectric-filled slotted coaxial waveguide of arbitrary cross section.

（任意の断面をもつ誘電体装荷スロット付き同軸導波管からの電磁回折を計算するための簡単な手順を要約する）

□ This paper summarizes the results of the test program.

（本論文では，試験プログラムの結果を要約する）

□ We have summarized a number of important concepts that are essential for establishing a high probability of detecting CMOS IC defects.

（ＣＭＯＳ　ＩＣ欠陥の高検出率を確立するのに不可欠な，多くの重要な概念を要約する）

□ Table 2 summarizes the measured harmonic distortions togeter with the SPICE2 simulation results.

（図２にSPICE2シミュレーション結果とともに，測定された高調波歪みをまとめている）

ポイント☞［類］abstract　　n:summarization 要約，summary 要約

suppose

(supposed, supposed, supposing)

〈他〉（事を）仮定する

be supposed to do:～することになっている，～することを要求されている

suppose that節:～ということを仮定する

☐ The sampled data are supposed to satisfy the sampling theorem.
(そのサンプル値データはサンプリング定理を満たすことを要求されている)

☐ Suppose that the layout element under consideration is a MOSFET shown in Fig.3.
(検討中のレイアウト要素は，図３に示すMOSFETだということを仮定する)

☐ Let us now suppose that the position of the manipulator is computed from the joint angles.
(さて,マニピュレータの位置はジョイントの角度から計算されると仮定しよう)

ポイント☞［類］assume, presume, hypothesize　　n:supposition 仮定
☞「多少とも根拠があって仮定する」のニュアンスがある．命令形でしばしば使われる．進行形は不可．

suppress
(suppressed, suppressed, suppressing)
〈他〉（振動・雑音などを）抑制する，除去する，低減する

☐ Low-order harmonics can be suppressed applying a PWM control method.
(低次高調波は，ＰＷＭ制御法を適用することで抑制できる)

☐ The nonlinearity suppresses the natural polarisation rotation phenomenon of the second-order modes.
(その非線形性のため，二次モードの固有偏波回転現象は抑制される)

☐ Permanent magnets are used to suppress saturation in the iron and extend the linear operating range.
(永久磁石は，鉄の飽和を抑制し，線形動作範囲を拡大するのに用いられる)

☐ To suppress the burst noise, hard limiter as an amplitude limiter is adopted in this system.
(バーストノイズを低減するために,振幅リミッタとしてのハードリミッタを

このシステムで採用する）

☐ An estimate of the average power specturm of the noise can be useful in suppressing the noise in order to enhance flaw detection.
（雑音の平均パワースペクトルの推定値は，探傷を向上させるために雑音を低減するのに役立つ）

☐ The phase error can be reduced by suppressing the amplitude of the harmonics.
（位相誤差は，高調波の振幅を抑制することで低減できる）

☐ Such composite doping is found to be very effective in suppressing the degradation of magnetic properties caused by the oxidation of the RE elements in the films.
（そのような複合ドーピングは，膜のＲＥ要素の酸化に起因する磁気的性質の劣化を抑制する際に，非常に効果的であるということがわかっている）

ポイント☞［類］restrain　　n:suppression 抑制

switch

(switched, switched, switching)

〈他〉（装置などの）スイッチを切り換える
switch A on/on A：Aのスイッチを入れる
switch A off/off A：スイッチを切る

☐ A glow lamp is switched on.
（グロー電球のスイッチを入れる）

☐ When an adaptive controller is first switched on, it is difficult to determine initial values for estimated model parameters.
（適応制御器が最初にスイッチを入れられると，推定されたモデルパラメータの初期値を求めることは困難となる）

☐ The memory does not lose stored bits after power is switched off.
（電力が切られた後でも，メモリは格納されたビットを失わない）

☐ A computerized control unit switched the laser on and off as required and controlled the speed and direction of the table.
(コンピュータ化制御装置は,必要に応じてレーザのスイッチをオン/オフにし，テーブルの速度と方向を制御した)

ポイント☞名詞としては「スイッチ・開閉器」. switching は切り換え. analog switch(アナログスイッチ), packet switched network(パケット交換網), circuit switching(回線交換)

synchronize

(synchronized, synchronized, synchronizing)

〈他〉（回路・信号などを)同期する，同調する

synchronize A with B：AをBと同期させる

☐ All data acquisition modules are synchronized with the sampling rate.
(データ収集モジュールはすべて，標本化周波数と同期する)

☐ The oscilating frequency is synchronized with the output signal in a lock-in amplifier to improve the measurement sensitivity.
(発振周波数は,測定感度を向上させるためにロックイン増幅器の出力信号と同期する)

☐ The crystal synchronizes the oscillator at the fundamental frequency of the crystal or its harmonics.
(その水晶は水晶の基本振動数，あるいはその高調波で発振器を同期させる)

☐ The arrival of data via the network is used to synchronize the processors.
(ネットワークを経由して到来したデータは,プロセッサを同期化させるのに用いられる)

ポイント☞n:synchronization 同期　adj:synchronous 同期の
☞bit synchronization(ビット同期), synchronous motor(同期電動機), synchronous generator(同期発電機)

synthesize

(synthesized,synthesized,synthesizing)

〈他〉（信号・回路などを）合成する

☐ A data path is synthesized from a data flow graph representation by replacing every node with a functionally equivalent circuit module.
（データ経路は，すべてのノードを機能的な等価回路モジュールと取り替えることでデータフローグラフ表現から合成される）

☐ Module generators are programs that synthesize module layouts from given input parameters.
（モジュールジェネレータは，ある特定の入力パラメータからモジュールレイアウトを合成するプログラムである）

☐ A simple method to synthesize the needed sinusoidal waveforms has been adopted.
（必要な正弦波形を合成する簡単な方法が採用された）

☐ The updated weights W_1 and W_2,which are needed to synthesize V_x, are calculated.
（V_xを合成するのに必要とされる更新された荷重W_1，W_2を計算する）

☐ The circuit to be synthesized is usually expressed as a flow table, a form similar to a truth table.
（合成される回路は通常，真理値表に類似した形式のフローテーブルとして表される）

ポイント☞n:synthesis 合成　adj:synthetic 合成の
☞synthesizer はシンセサイザー． speech synthesis (音声合成), synthetic resin(合成樹脂), synthetic compound(合成化合物), synthetic fiber(合成繊維)

take

(took, taken, taking)

〈他〉 (時間を)費やす

(注意・決心・見方などを)する

(データなどを)とる

(形状・性質などを)とる

take account of A＝take A into account：Aを考慮に入れる

take advantage of A：Aを利用する

take care of A：Aに注意する

take part in A：Aに加わる，直面する，関係する

take place：起こる，行われる

take up A：Aを取り上げる

☐ Data were taken for films of 1 μm in thickness.
(厚さが1 μmの膜のデータをとった)

☐ To simplify treatment, the transformer turns ratio is taken to be 1:1:1.
(取扱いを容易にするために，変圧器の巻数比を1:1:1とする)

☐ Extra care was taken to avoid any electronic pickups, especially at frequencies higher than 90 MHz.
(とりわけ90 MHz以上の周波数で，電子ピックアップを避けるために，特別な注意を払った)

☐ It takes about 1 minute to simulate a single frame of video compression on a high end workstaion.
(ハイエンドワークステーションのビデオ圧縮の単一フレームをシミュレートするのに，およそ1分かかる)

☐ These interferometers take various forms including Mach-Zehnder, Sagnac and Fabry-Perot.
（これらの干渉計は，Mach-Zehnder，Sagnac，Fabry-Perotを含むさまざまな形式をとる）

☐ According to experimental and numerical results, the Fresnel zone is sufficient to take account of the region of power line interference.
（実験結果と数値結果によれば，このフレネル帯は電力線妨害の範囲を十分に考慮できる）

☐ The parabolic dispersion relation does not take into account the existence of the valence band in the SiO_2.
（その放物分散関係では，SiO_2での価電子帯の存在を考慮に入れていない）

☐ When designing a tracking system, several aspects must be taken into account.
（追跡システムを設計する場合，いくつかの側面を考慮しなければならない）

☐ The pipeline cascade implementation takes advantage of the FFT algorithm.
（そのパイプラインカスケードの実現では，ＦＦＴアルゴリズムを利用している）

☐ Algorithmic CAD tools are needed to take care of these optimizations.
（アルゴリズミックＣＡＤツールでは，これらの最適化に留意する必要がある）

☐ As the contacts in common use are connected to lead lines, it should be assumed that lines take part in the discharge phenomena.
（通常使われている接点はリード線に接続されているので，線は放電現象に直面するものと仮定するべきである）

☐ Using a low viscosity adhesive, bonding should take place under pressure.
（低粘度接着剤を使えば，圧力のかかった状態で接合が起こるはずである）

☐ Many applications are taking place in centimeter waves,so millimeter waves (30

GHz< F <300 GHz) are more frequently used in radar and telecommunication systems.
(多くの応用例はセンチメートル波で行われつつあり，そのためミリメートル波(30 GHz< F <300 GHz)は，レーダや通信システムでより頻繁に用いられている)

□ The variance of δ is most easily determined in the absence of a signal, and so we take up this case first.
(δ の分散は信号なしで非常に容易に決定できるので，まず最初にこの場合を取り上げる)

□ To start with, we take up the nonlinear transfer function of a bipolar transistor.
(まず初めに，バイポーラトランジスタの非線形伝達関数を取り上げる)

ポイント☞take は多くの熟語を作る．他の熟語をいくつか取り上げよう．take it for granted that～(～を当然と考える)，take off(離陸する)，take notice of(～に注意する)，take A for B(AをBと間違える)，take time(時間がかかる)，take apart(分解する)，take away(取り去る)

tend

(tended, tended, tending)

〈自〉 (物・事が)～する傾向がある
tend to do ～する傾向がある，～しがちである
(物・事がある結果・状態に)向かう，至る

□ The efficiency-bandwidth product tends to remain constant.
(その効率・帯域幅積は一定にとどまる傾向がある)

□ If rain is present in the form of relatively small drops, it is seen that the ratio would tend to increase.
(雨が比較的小さなしずくの形であれば，この割合は増加する傾向にあるということがわかる)

☐ Certain combinations of the number of slots and the number of turns tend to minimize the error in the angular variation of the magnetomotive force(MMF).
（スロット数と巻数をある特定の組合わせをすると，起磁力（MMF）の角変化の誤差を最小とする傾向が見られる）

☐ This result does not tend towards the accurate value.
（この結果は，正確な値に至らない）

ポイント☞n:tendency 傾向

☞tend to do では進行形は不可．tend to＝have a tendency to

test

(tested, tested, testing)

〈他〉（物・事を）試験する

☐ The device was tested with the experimental set-up shown in Fig.6.
（装置は図6に示してある実験装置で試験した）

☐ Can this chip be designed and tested within a reasonable length of time？
（このチップは妥当な時間内で設計，および試験が可能だろうか）

☐ The robustness of the designed compensator can be tested in many ways.
（その設計された補償器のロバスト性は，いろいろな方法で試験できる）

☐ This circuit can be tested by measuring the input and output voltage amplitudes for a variety of frequencies.
（この回路は，さまざまな周波数に対して入力と出力の電圧振幅を測定することで試験できる）

☐ We have tested these algorithm on various images.
（いろいろな画像に対してこれらのアルゴリズムを試験した）

☐ A simple voltmeter was sufficient to test the correctness of each SSI chip.
（各*SSI*チップの正確さを試験するには，簡単な電圧計で十分である）

☐ Open-circuit magnetic measurements are used for testing permanent magnets.
(開路磁気測定値は，永久磁石の試験に使われる)

☐ The accuracy of these formulas is estimated to be better than 10% for all the cases tested, as demonstrated on an example in Section Ⅳ.
(これらの公式の精度は，第4節での一例で例証されたように，試験されたすべての場合に対して10%以上優れていると推定される)

ポイント☞名詞としては「試験・検査」．testing も試験．tester は試験器・テスター．nondestructive testing(非破壊試験), acceptance testing(受け入れ試験)

transduce

(transduced, transduced, transducing)

〈他〉（エネルギー・信号などを）変換する
transduce A into B：AをBに変換する

☐ Voice input from the operator is transduced into an electrical signal using a microphone.
(オペレータからの音声入力はマイクロホンを用いて電気信号に変換される)

☐ Light is transduced into an electrical signal via the photoreceptors.
(光は光受容体によって電気信号に変換される)

☐ We use the circuit shown in Fig.2 to transduce light into an electrical signal.
(光を電気信号に変換するために，図2に示す回路を用いる)

ポイント☞［類］convert, transform, translate
☞transducer は変換器・トランスジューサ．ultrasonic transducer(超音波変換器)

transfer

(transferred, transferred, transferring)

〈他〉（データなどを）転送する，（物を）移動させる，移す

☐ The data are transferred by the DMA (direct memory access) operation through each data bus.
（データは各データバスを通してDMA（直接メモリアクセス）動作によって転送される）

☐ The amplitude and phase data are recorded on floppy disc and later transferred to a computer for processing.
（振幅および位相データはフロッピーディスクに記録され，その後処理するためにコンピュータへ転送される）

☐ The electrodes were transferred directly to the electrochemical cell.
（電極は電気化学的電池へ直接移送される）

☐ This manufacturing equipment is employed throughout the fabrication process, so that it may be easily transferred to a commercial production line.
（この製造装置は製造工程全体を通して使用され，そのためそれを商業用生産ラインに容易に移送できる）

☐ Data is collected continuously in the RTU and periodically transferred to the computer in the central control facility via modems or radio links.
（データはRTUに間断なく収集され，モデムか無線リンクによって中央制御施設内のコンピュータへ定期的に転送される）

☐ Waveforms are recorded on a digital oscilloscope and transferred to a personal computer via a GPIB interface.
（波形はディジタルオシロスコープに記録され，GPIBインタフェースによってパーソナルコンピュータへ転送される）

☐ The electrostatic reaction transfers the force to the crystal lattice charges, causing a mechanical force on QR.
（静電気反作用は力を結晶格子電荷に伝え，その結果QRに機械力をもたらす）

ポイント☞［類］move, shift, migrate, transport
☞「一つの場所か別の場所へ移す」のニュアンスがある．名詞では「転送・移動」. data transfer(データ転送),file transfer(ファイル転送)

transform
(transformed, transformed, transforming)
〈他〉（物を）変換する，変形する
transform A into（to）B：AをBに変換する

□ The phase information of the imaging element is transformed into amplitude information on the screen.
(その画像素子の位相情報はスクリーン上で振幅情報に変換される)

□ The discrete Fourier transform transforms the signal to the frequency domain.
(その離散フーリエ変換は，信号を周波数領域に変換する)

□ It is necessary to transform this digital signal to its analog signal in order to drive amplifiers and speakers.
(増幅器とスピーカを駆動させるためには,このディジタル信号をアナログ信号に変換する必要がある)

□ The transformed boundary conditions are applied at the interface between the two dielectrics.
(その変形境界条件は，二つの誘電体間の界面に適用される)

ポイント☞［類］convert, translate, transduce　n:transformation 変換
☞「構造・性質・機能などを全く別の形態に変換する」のニュアンスがある．名詞としては「変換」. transformer は変圧器・トランス．coordinate transformation(座標変換),Fourier transform(フーリエ変換)

transmit
(transmitted, transmitted, transmitting)
〈他〉（光・熱などを）伝える

(電波・信号・データなどを)送信する，伝送する)
transmit A to B：AをBへ送る

☐ Light is transmitted from the bulkhead to the lens via optical fibers.
(光は光ファイバを介してバルクヘッドからレンズへ送られる)

☐ The encoded data derived from the hidden layer units are transmitted to the decoder.
(隠れ層ユニットから導出された符号化データは，復号器へ伝送される)

☐ If there is air in the slot, the energy is not transmitted.
(スロット内に空気が存在すれば，エネルギーは伝わらない)

☐ This system operates in the 400-Mbit/s FSK single filter detection scheme, which can transmit an HDTV signal.
(このシステムは400-Mbit/s FSK単一フィルタ検出方式で動作し，HDTV信号を送信することができる)

☐ For example,in the case of high-definition television (HDTV),a channel to transmit all the HDTV image information would require immense bandwidth capacity.
(例えば，高精細度テレビジョン(HDTV)の場合，ＨＤＴＶ画像情報のすべてを伝送する１チャンネルは多大な帯域幅能力を必要とすることになるだろう)

ポイント☞［反］receive(受信する)　［類］send　n.transmission 伝送・送信.
☞transmitter は送信機．data transmission(データ伝送)，transmission line(伝送路)

travel
(traveled, traveled, traveling)
〈自〉(光・音などが)伝わる，移動する

☐ The holes will travel across the doped drift region of width l_d.

(正孔は幅がl_dのドープされたドリフト領域を横切って移動する)

☐ As the plasma sheet travels along the electrodes the circuit inductance changes.
(プラズマシートが電極に沿って伝わるため,回路のインダクタンスは変化する)

☐ This light travels down the optical fiber towards the probe, which is held close to the specimen surface.
(この光は光ファイバを伝わってプローブへと向かうが,このプローブは試料表面の近くに保持されている)

☐ We consider a Rayleigh wave traveling in the X_1 direction as shown in Fig.1.
(図1に示されているように,X_1方向に伝わるレイリー波について考察する)

☐ The transducer responds to an acoustic beam travelling in the opposite direction.
(このトランスジューサは,反対方向に伝わる音響ビームに応答する)

☐ In the 1000-kHz AM radio wave region,electromagnetic fields in free space traveling at the speed of light have wavelengths on the order of 300 m.
(1000-kHz AM無線領域の場合,光の速度で伝わる自由空間における電磁場は,300 mのオーダーの波長を有する)

ポイント☞［類］propagate, conduct
☞名詞としては「移動」.

treat

(treated, treated, treating)

〈他〉(物・事を)扱う,処理する,論じる
treat A with B:AをBで処理する
treat A as B:AをBとみなす

☐ The problem cannot be accurately treated with analytical means.
(その問題は解析的方法では正確に扱うことはできない)

☐ The former will be treated by the finite element method and the latter by the boundary element method.
(前者は有限要素法で，そして後者は境界要素法によって取り扱われる)

☐ The important case of a plasma between the electrodes is not treated here.
(電極間プラズマについての重要な事例はここでは扱わない)

☐ The permeability of region 1 can be treated as constant and isotropic.
(領域１の透磁率は一定で等方性だとみなせる)

☐ In Section Ⅳ we treat the problem of determining an optimal system.
(第４節では，最適システムを決定する問題について論じる)

☐ Since it is common practice in system engineering to treat nonlinear systems as linear time-varying systems, the identification of linear time-varying systems should be an important topic in system identification.
(非線形システムを線形時変システムとして扱うことは，システム工学では一般的な方法なので，線形時変システムの同定は，システム同定における重要な論題となるはずである)

☐ The microstrip circuit treated as a three-layer model is illustrated in Fig.1.
(三層モデルとみなせるマイクロストリップ回路は図１に示している)

ポイント☞［類］handle, process　　n:treatment 取り扱い・処理
☞「特定の方針・考え・見地のもとで扱う」のニュアンスがある．

try

(tried, tried, trying)

〈他〉（物・事を）試す，試みる
try to do:～しようと試みる

☐ We tried a single layer of 0.5-μm-thick MP2400 resist on 10-μm oxide.
(10-μm酸化物上に0.5-μm厚のMP2400レジストの単層を試みた)

☐ We have tried to give an overview of the current status of dc and RF SQUIDs.
(直流SQUIDとRF SQUIDの現状の概要を示すことを試みた)

☐ For this reason, we will try to design and to develop a polarization- independent attenuator.
(この理由で，偏波に依存しない減衰器の設計と開発を試みる)

☐ According to the analyses on the ac resistivity,we tried to analyze the elements of the power loss for Mn-Zn ferrites in order to improve the power loss characteristics.
(交流抵抗率の解析に従って，電力損失特性を向上させるために，Mn-Znフェライトの電力損失成分の解析を試みた)

ポイント☞［類］attempt
☞名詞としては「試み」.

tune
(tuned, tuned, tuning)
〈他〉(ある値に)調整する
(受信機・レーザなどを)同調させる
tune A to B：AをBに合わせる，同調させる

☐ The DX2070 can be tuned via 8-b microprocessor control.
(DX2070はマイクロプロセッサ制御で調整できる)

☐ The wavelength is tuned from 948.8 to 958.9nm when the heater current increased from 0 to 174 mA.
(ヒータの電流が0から174 mAに増加する時に，波長は948.8～958.9nmに同調される)

☐ The gain can be tuned by ±4.5 dB by adjusting a control voltage between 0.5 and 2.5 V.
(利得は，0.5～2.5 V間の制御電圧を調節することで，±4.5 dBだけ調整できる)

☐ The center wavelength of the LD can be tuned to the reference value by a temperature adjustment.
（このLDの中心波長は，温度調節によって基準値に合わせることができる）

☐ Series reactance X_s is added in order to tune to the resonance imaginary components of the impedances.
（インピーダンスの共振虚成分を調整するために，直列リアクタンスX_sを加える）

☐ Since the error of the filter characteristics in this method is only observed in a frequency shift, there is no need to tune the Q factor and the error can easily be tuned by conventional manners.
（本方法におけるフィルタ特性の誤差は周波数偏移として観測されるだけなので，Q値を調整する必要はなく，誤差は従来のやり方で容易に調整できる）

☐ An argon laser, tuned to λ =514 nm, is used as the light source.
（λ =514に同調されたアルゴンレーザは，光源として使われる）

ポイント☞［類］regulate, adjust
☞名詞としては「同調」．tuning も同調．self-tuning regulator(自己同調レギュレーター)

U

understand
(understood, understood, understanding)
〈他〉（事を）理解する

☐ Since the flicker noise is not well understood theoretically, it must be determined empirically.

(フリッカ雑音は理論的にはあまりよく理解されていないので，実験に基づいて決めなければならない)

☐ It is easily understood that the minimum BER can be obtained when *v* is chosen so that *R* is maximized.
(*R*が最大となるように*v*を選ぶと，最小のBERが得られるということは容易に理解できる)

☐ In order to discuss its performance as a detector, it is essential to understand the photocurrent in a p-i(MQW)-n structure.
(検出器としてのその性能を論じるには，p-i(MQW)-n構造での光電流を理解することが不可欠である)

ポイント ☞［類］comprehend　n:understanding 理解
☞「表面的な意味だけでなく，その意味内容を十分に理解する」のニュアンスがある．image understanding(画像理解)

use

(used, used, using)

〈他〉（物・事を）使う，使用する，利用する

☐ Pulse echo techniques are widely used for the ultrasonic inspection of materials.
(パルス・エコー法は，材料の超音波検査に広く使われている)

☐ Fiber optic cables are being used for data highways.
(光ファイバケーブルはデータハイウェイに使用されてきている)

☐ The technique of ion implantation has been widely used in semiconductor device technology.
(このイオン注入法は，半導体デバイス技術で幅広く使われている)

☐ There are several types of neural networks that can be used in control systems:the multilayer perceptron,Kohonen's self-organising map, the Hopfield network, the Boltzmann machine,etc.

(制御システムに用いることができるニューラルネットワークは数種類ある．すなわち，多層パーセプトロン，コーネンの自己組織化写像，ホップフィールド・ネットワーク，ボルツマンマシン，その他である)

□ A rechargeable battery is used as the power supply for the active network.
(蓄電池は能動回路網用の電源として用いられる)

□ A novel device is described which can be used as an optical fibre sensor.
(光ファイバセンサとして使用することが可能な，新しい装置について述べる)

□ Mobile satellite communication systems are usually used at low data rates.
(移動衛星通信システムは，低データ転送速度で通常使用される)

□ This antenna uses a synthetic aperture technique.
(このアンテナでは，合成開口手法を使用している)

□ We use a pulsed light emitting diode as the stroboscope.
(パルス発光ダイオードをストロボスコープとして用いる)

□ The private communication protocol uses the data link layer of the ISO/OSI reference model.
(専用通信プロトコルでは，ISO/OSI参照モデルのデータリンク層を用いる)

□ Two widely used digital image compression techniques are predictive coding and transform coding.
(広く用いられている２種類のディジタル画像圧縮法は，予測符号化と変換符号化である)

ポイント☞［類］employ, utilize, make use of, take advantage of　　adj:useful 有用な，useless 役に立たない
☞名詞としては「使用・利用」. be in use(使用されている), of use(役に立つ)

utilize

(utilized, utilized, utilizing)

〈他〉（物・事を）利用する，役立たせる

□ This device can be utifized for the DPSK demodulator in digital communication.
（この装置はディジタル通信のDPSK復調器に利用できる）

□ Commonly, Stark modulation is utilised in a microwave spectrometer by means of an applied alternating electric field, *E*, generated by a metal electrode in the gas cell.
（一般に，シュタルク変調は，気体電池内の金属電極によって発生する印加交番電場*E*を用いて，マイクロ波分光計で利用される）

□ The motor does not utilize a conventional commutation sensor or a brushless tachometer for determining the absolute rotor position.
（この電動機では，回転子の絶対位置を決定するために，従来の整流センサやブラシレス回転速度計を利用しない）

□ These converters utilize multistage noise shaping modulation techniques which can be implemented with VLSI MOS technology.
（これらの変換器は，VLSI MOS技術で実現できる多段雑音整形変調法を利用している）

□ Modern energy management systems (EMS) utilized for electric utility control and monitoring are required to provide real-time security assessment.
（電力会社における制御と監視に利用される現代のエネルギー管理システム(EMS)は，実時間セキュリティ評価をもたらすのに必要とされる）

ポイント☞［類］employ, use, make use of, take advantage of　n:utilization 利用
☞「ある物を実用的に利用する」のニュアンスがある．

v

vary

(varied, varied, varying)

〈他〉（物・事を）変える，変化させる

〈自〉（物・事が）変わる，変化する

vary with A：Aによって変わる，Aにつれて変わる

☐ The propagation velocity can be varied by changing the applied magnetic field.
（その伝搬速度は印加磁場を変化させることで変えることができる）

☐ Optical transmission between the input and the output channel waveguides can be varied by changing the refractive index distribution in the connecting waveguide.
（入力および出力チャネル導波管の間の光伝送は，コネクティング導波管の屈折率分布を変化させることで変えることができる）

☐ With the VCO, the control voltage can be varied over a broad range, and the oscillation frequency varies in proportion to the control voltage.
（VCOによって，制御電圧を広い範囲にわたって変えることができ，振動数は制御電圧に比例して変化する）

☐ We vary the summation indices (m) in Eqs.(5) and (7) until the solution converges.
（解が収束するまで，方程式(5)，(7)の総和指数(m)を変える）

☐ We varied the resonator lenght between 0.6 and 4 cm.
（共振子の長さを0.8から4 cmの範囲で変えた）

☐ The thermal conductivity of gallium arsenide varies considerably with

temperature.
(ガリウムひ素の熱伝導率は温度につれてかなり変化する)

☐ Speckle noise of the SIR-B images was so strong that image intensity varies greatly.
(SIR-B画像のスペックル雑音は非常に強いので,画像の輝度は大きく変化する)

☐ The resonant frequency varies from 6.9 to 7.6 MHz.
(その共振周波数は6.9 MHzから7.6 MHzである)

☐ The difference between the peak and average temperature varies for different operating conditions and devices.
(最高温度と平均温度の差は,さまざまな運転条件と装置に対応して変わる)

☐ Using Equation (26), the reflection coefficient varying with respect to the number of the cracks can be evaluated.
(方程式(26)を使えば,亀裂の数に対して変化する反射係数を求めることができる)

ポイント☞［類］change, alter, convert　n:variation 変化　adj:various さまざまの,variable 変わりやすい
☞「構造・性質などがしだいに変化する」のニュアンスがある. variable には変数の意味もある. dependent variable(従属変数)

verify
(verified, verified, verifying)
〈他〉(事を)立証する,実証する,検証する,確かめる
verify that節:〜ということを確かめる

☐ This was verified experimentally over a wide range of dither frequencies.
(このことを広い範囲のディザ周波数にわたって実験で立証した)

☐ The concept has been verified experimentally using a 1.2 μm CMOS process.

（その概念を，1.2 μm CMOSプロセスを用いて実験で立証した）

☐ This can be verified by calculating projections of Poincaré maps.
（このことはポアンカレ写像の射影を計算することで立証できる）

☐ Analytical results are verified with both frequency-domain and time-domain noise measurements.
（解析結果を周波数領域と時間領域両方の雑音測定によって実証する）

☐ One can verify in detail that the charge conservation condition is satisfied.
（この電荷保存条件が満たされるということを，詳細に確かめることができる）

☐ Several examples are given to verify the accuracy and advantages of the proposed method.
（提案された方法の精度と利点を検証するための例を与える）

☐ A large number of examples were tested to verify the convergence properties of the iterative linear programming algorithm.
（反復線形計画アルゴリズムの収束性を検証するために，多くの事例を試験した）

☐ In this section we describe the experiments that were performed to verify the theroretical predictions.
（本節では，理論的予測値を検証するために実施された実験について述べる）

ポイント☞［類］demonstrate, confirm, ascertain, make sure　　n:verification立証
☞「検討・比較によって事実を立証する」のニュアンスがある．

view

(viewed, viewed, viewing)

〈他〉（物を）見る，観察する
view A as B：AをBとみなす

☐ Until recently, communications and computers were viewed as separate disciplines.
(最近まで，通信とコンピュータは別々の分野だとみなされていた)

☐ A robot can be viewed as a physical mechanism for performing work.
(ロボットは作業を行うための物理機構とみなすことができる)

☐ The basic circuit of Fig.2 can be viewed from two different angles.
(図２の基本回路は二つの異なる角度から見ることができる)

☐ In order to view the surface structure and obtain detailed images of the topology, the films were analysed using SEM.
(表面構造を観察し，配列の詳細な画像を得るために，膜をＳＥＭを用いて分析した)

ポイント☞［類］see, regard, observe
☞“view A as B"において，as の後には名詞・形容詞がくる．consider と混同して asを to be とはしないこと.名詞としては「視野・眺め」.

W

work

(worked, worked, working)

〈自〉（機械などが）動作する，作動する，機能する
（計画・方法などが）うまくいく

☐ The adaptive DCT works poorly at low bit rates.
(適応DCTは低ビットレートではあまりうまくいかない)

☐ This method works well for detection of average flow rate only.

(この方法は，平均流量の検出にのみ具合がよい)

☐ From Fig.4 it can be seen that a 16 state Viterbi equalizer can work with a time dispersion to around 16 micro seconds.
(図 4 から, 16状態ヴィテルビ等化器は約16マイクロ秒までの時間分散で動作できるということがわかる)

☐ The CCD image sensor works at 12 V power supply, because a charge detector needs 12 V power supply.
(このCCDイメージセンサは12V電源で機能する．というのは，電荷検出器は12V電源を必要とするからである)

ポイント☞［類］operate, run, drive, function, manipulate　adj:workable 使用可能な・実行可能な
☞名詞としては「作業・仕事・工作物」.

write
(wrote, written, writing)
〈他〉（数式・プログラムなどを）書く

☐ Maxwell's equation for the two-dimensional magnetostatic problem can be written as follows.
(二次元静磁気問題に対するマクスウェル方程式は, 次のように書くことができる)

☐ A compiler was written to generate the microcode for the communication sequencer.
(通信シーケンサ用のマイクロコードを生成するために，コンパイラを書いた)

☐ According to (1.7),(5.1) can be written in the form (1.1).
((1.7)によれば，(5.1)は(1.1)の形で書くことができる)

☐ A number of efficient algorithms have been written to implement the Fast

Fourier Transform (FET).
(高速フーリエ変換(FET)を実行するために，多くの効率的なアルゴリズムが書かれてきた)

□ These values were written on paper and plotted by hand.
(これらの値は紙に書かれ，手書きでプロットされた)

□ We first write $H(s,t)$ in the fractional form (2).
(最初に分数形式（2）で$H(s,t)$を書く)

□ Due to (4.3) we can write down that $y(t)=Q(t)+\phi(t)$.
((4.3) のために，$y(t)=Q(t)+\phi(t)$と書くことができる)

□ We will write the propagation constants β_j in terms of the effective refractive indexes n_j of the fiber waveguide at ω_j; that is, $\beta_j=\omega_j n_j/c$.
(ファイバ導波管の有効屈折率n_jによって，伝搬定数β_jを書き表す．すなわち，$\beta_j=\omega_j n_j/c$となる)

□ We can easily write mathematical expressions describing the current and charge distributions for this simple model.
(この簡単なモデルのための電流分布と電荷分布を記述する数式は，容易に書ける)

ポイント☞n:writing 書き込み

☞write protection(書き込み保護)，writing head(書き込みヘッド)

Y

yield

(yielded, yielded, yielding)

〈他〉（物・事を）生じる，もたらす
（物・事が）得られる

☐ The optimal solution of the linear programming yield the structure of the optimal control.
（この線形計画法の最適解により，最適制御の構造が得られる）

☐ Convolution of frequency-domain representations yields the frequency spectrum of a sampled signal.
（周波数領域表現の畳込みにより，標本化信号の周波数スペクトルが得られる）

☐ Numerical results for two-dimensional structures show that the proposed method yields a faster convergence rate than the original ones.
（二次元構造に対する数値結果により，提案された方法は元の方法よりも速い収束速度をもたらすということを示す）

☐ The model is too simple to yield accurate measurements and can only be used as a guideline in design.
（そのモデルはあまりにも単純なものなので，正確な測定値が得られず，設計における指針としてのみ使うことができる）

ポイント☞［類］produce, generage, create, bring, give rise to
☞名詞としては「生産高・歩留り」.

著　者　略　歴

宮野　晃（みやの・あきら）

1959年　北海道生まれ

神奈川大学で電気工学を専攻．国外の科学技術文献を読む必要性から，技術英語に興味を抱き，在学中から技術論文の翻訳を手掛ける．

卒業後，商社を経て，現在は科学技術関係のデータベース作成に従事．技術英語の教育にも関心をもち，月刊誌「電子材料」，「自動化技術」に技術英語の講座を連載する．

著書に「技術英語らくらく表現法」，「はじめての技術英語」，「技術英語よく使う基本語句と表現」他がある．

理系，文系の両分野にわたった学際領域の研究をライフワークにしている．

北海道在住．

電気・電子を説明する英語

2011年8月16日　第1版第1刷発行

著　　者　宮野　晃

発 行 者　森北博巳

発 行 所　**森北出版株式会社**

東京都千代田区富士見1-4-11（〒102-0071）

電話 03-3265-8341／FAX 03-3264-8709

http://www.morikita.co.jp/

日本書籍出版協会・自然科学書協会・工学書協会　会員

JCOPY ＜(社) 出版者著作権管理機構　委託出版物＞

落丁・乱丁本はお取替えいたします　　印刷・製本/藤原印刷

Printed in Japan／ISBN978-4-627-94601-9